An Expedition of Discovery into the Interior of Africa through the hitherto undescribed countries of the Great Namaquas, Boschmans and Hill Damaras, etc. [With plates.]

James Edward Alexander

An Expedition of Discovery into the Interior of Africa through the hitherto undescribed countries of the Great Namaquas, Boschmans and Hill Damaras, etc. [With plates.]
Alexander, James Edward
British Library, Historical Print Editions
British Library
1838
2 vol. ; 8°.
1047.i.13.

The BiblioLife Network

This project was made possible in part by the BiblioLife Network (BLN), a project aimed at addressing some of the huge challenges facing book preservationists around the world. The BLN includes libraries, library networks, archives, subject matter experts, online communities and library service providers. We believe every book ever published should be available as a high-quality print reproduction; printed on- demand anywhere in the world. This insures the ongoing accessibility of the content and helps generate sustainable revenue for the libraries and organizations that work to preserve these important materials.

The following book is in the "public domain" and represents an authentic reproduction of the text as printed by the original publisher. While we have attempted to accurately maintain the integrity of the original work, there are sometimes problems with the original book or micro-film from which the books were digitized. This can result in minor errors in reproduction. Possible imperfections include missing and blurred pages, poor pictures, markings and other reproduction issues beyond our control. Because this work is culturally important, we have made it available as part of our commitment to protecting, preserving, and promoting the world's literature.

GUIDE TO FOLD-OUTS, MAPS and OVERSIZED IMAGES

In an online database, page images do not need to conform to the size restrictions found in a printed book. When converting these images back into a printed bound book, the page sizes are standardized in ways that maintain the detail of the original. For large images, such as fold-out maps, the original page image is split into two or more pages.

Guidelines used to determine the split of oversize pages:

Some images are split vertically; large images require vertical and horizontal splits.
For horizontal splits, the content is split left to right.
For vertical splits, the content is split from top to bottom.
For both vertical and horizontal splits, the image is processed from top left to bottom right.

UNUMA MOUNTAINS.

Etched by Wm Leaf

Published by Henry Colburn, 13 Great Marlborough Street, 1842

AN

EXPEDITION

OF

DISCOVERY

INTO THE

INTERIOR OF AFRICA,

THROUGH THE HITHERTO UNDESCRIBED

COUNTRIES OF THE GREAT NAMAQUAS, BOSCHMANS,
AND HILL DAMARAS.

PERFORMED

Under the auspices of Her Majesty's Government,

AND

THE ROYAL GEOGRAPHICAL SOCIETY;

AND CONDUCTED BY

SIR JAMES EDWARD ALEXANDER, K.L.S.

CAPTAIN IN THE BRITISH, LT.-COLONEL IN THE PORTUGUESE, SERVICE,
F.R.G.S. AND R.A.S., ETC.

IN TWO VOLUMES.

VOL. I.

LONDON:
HENRY COLBURN, PUBLISHER,
GREAT MARLBOROUGH STREET.

———

1838.

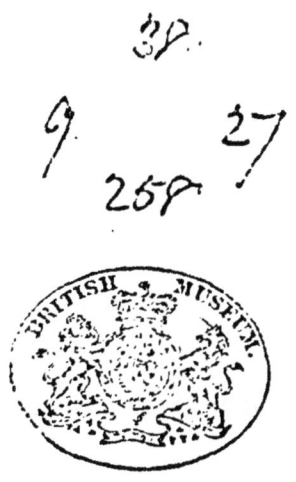

PRINTED BY WILLIAM WILCOCKSON, ROLLS BUILDINGS, FETTER LANE.

TO

THE RIGHT HONOURABLE

THE LORD GLENELG,

HER MAJESTY'S PRINCIPAL SECRETARY OF STATE FOR THE
COLONIES, ETC.

THESE VOLUMES

ARE, BY PERMISSION, RESPECTFULLY INSCRIBED,

BY HIS LORDSHIP'S MOST OBEDIENT

AND VERY HUMBLE SERVANT,

THE AUTHOR.

INTRODUCTION.

" Avia Pieridum peragro loca, nullius ante
   Trita solo ; juvat integros accedere fontes,
   Atque haurire ; juvatque novos decerpere flores."

In the beginning of the year 1834, the Author of the following pages had volunteered from the depôt of the 42nd Royal Highlanders, in Scotland, to join the service companies of the regiment abroad, and was accordingly proceeding to the Mediterranean, when he had the honour to receive an invitation from the Royal Geographical Society, to perform an African Expedition of Discovery ; and, at the same time, he was recommended by that distinguished body to Government, as a fit person for conducting such an undertaking.

His humble services in the field before enemies of his country not being then required—moreover, he having had previous experience in voyaging and travelling, —having always had a strong desire to attempt to discover some of the secrets of the great and mysterious continent of Africa, and feeling that he could not rest satisfied till he had broken new ground there,—he readily consented to explore and report on certain regions of East Africa, from Delagoa Bay westwards, with a view to the extension of geographical knowledge and commerce.

In undertaking the conduct of an African expedition, he was actuated by no hope of pecuniary gain. He was perfectly independent in his circumstances; and his only hope was, that if he performed an expedition which should prove successful, and which might tend to promote trade, to civilize the native tribes that might be visited, and to extend a knowledge of our holy religion,—his expenses would be defrayed, and he should receive, in common with all other expedition-officers, some honorary reward for his labours, and for his having consented to exchange, during a certain period, civilized for savage life.

Having obtained leave of absence from the authorities at the Horse Guards, he proceeded at his own expense to Portugal, to collect information relating to Africa, and had an opportunity of being present with the contending parties in the field during the late civil war ; then,—recommended by the Admiralty to the care of Admiral Sir Patrick Campbell,—he sailed in the flag-ship Thalia, on a voyage of observation among the Colonies of Western Africa, and arrived at the Cape of Good Hope in the beginning of 1835.

It was now unexpectedly found that the whole of South Africa was in commotion; that the Zoolas had risen on the Portuguese settlement at Delagoa, the ultimate destination of the author, and had slain the Governor and some of his people ; that some of the native tribes were carrying on a war of extermination against each other, and that the Amakosa Kaffers had suddenly burst into the eastern province of the Cape Colony, and were carrying fire and sword through it.

It was evidently not the time for geographical research, and the Author, having a recommendation from Sir Herbert Taylor, his late Majesty's Private Secretary, for employment on the staff at the Cape, if an opportunity should offer for his serving in a military capacity there, was accordingly placed by General Sir Benjamin D'Urban, the Governor and Commander-in-chief, on the personal staff of his Excellency, to whom he became an aid-de-camp and private secretary.

After the conclusion of the Kaffer war, and when there seemed to be a lull in the interior, the Author prepared to carry the original purpose for which he had arrived in South Africa into effect, when Dr. A. Smith arrived at the Cape from an expedition to the north east, having passed over the ground behind Delagoa Bay, and having visited the Maquaina country, whither it was intended the Author should have gone. It became, therefore, necessary to change the original route.

It is remarkable that during the three centuries and a half which have elapsed since the celebrated Portuguese navigator, Bartolomeo Diaz, first doubled the "Cape of Storms," the progress of discovery should have advanced so slowly, that up to this day, the whole of the western region of Southern Africa, has hitherto remained comparatively a blank in our maps. The Great Fish River, supposed to extend upwards of three hundred miles from north to south, and said to receive, both from the eastward and from the westward, more than twenty tributaries, was only indicated by a dotted line; of the range and height of the mountains and elevated plains near it, no trace existed, and of their geological structure and general features, we were utterly ignorant.

Gordon in 1777, Paterson in 1778, Le Vailant in 1781, and Thompson in 1827, may have reached, *in this direction*, the southern bank of the Gariep, but, neither in the last nor present century, is it recorded, that any European traveller has crossed within four hundred miles of its mouth, to the northern bank of the Orange river.* But, where the spirit of geographical enter-prise had not yet reached, a nobler spirit had directed the steps of other wanderers, and, for nearly a quarter of a century, a few missionaries had, from time to time, endea-voured to spread the truths of the Gospel in the district of the Orange river; and one, the Rev. Mr. Schmelen, performed a journey of some extent, several years ago, through a part of Great Namaqua land.

* Barrow in 1797, Truter and Somerville in 1801, Lichtenstein in 1805, Burchell in 1809, Campbell in 1813, Thompson in 1827, Hume in 1834, with several others, and lastly, Dr. Smith in 1835-36, have crossed the Orange river, but that was between 23° and 24° east longitude, near its mid-course, or upwards of 400 miles from its mouth. A Dutch colonist, also, W. Van Reenen, went some distance up the Great Fish River, but his account was never published.

To penetrate further to the north, then, in this direction, and to endeavour to become acquainted with the Damaras, a nation inhabiting between the 21st and 24th parallels, and only known to us by report, seemed to be now the chief object of geographical research, since the recent journey of Dr. Smith had rendered it needless to proceed with the Delagoa expedition. Accordingly, after mature deliberation, with the authority of his Excellency Sir Benjamin D'Urban, and by the advice of those best qualified to give an opinion in the colony, the author determined to explore the country to the north of the Orange river, on the west coast, as the best means of promoting the great object for which he had left England, and carrying out the views of Her Majesty's Government, which contributed to his outfit, and of the Geographical Society, which has also advanced its funds to assist in covering the expenses of the undertaking.

In these volumes there will be found a

faithful account of an Expedition attended with many trials and troubles,—in which much hunger and thirst were endured— intense heat and severe cold ; to all which the traveller in Africa must make up his mind, if he intends to pass a considerable distance beyond the bounds of civilization. Throughout, the narrative is interspersed with anecdotes illustrative of African manners and of the chace, &c.* and it is confidently hoped, that those who honour this work with their perusal, will find in its details something that may not leave their curiosity ungratified.

The etchings and wood cuts are selected from many drawings made during the journey, and are accurately outlined by the artists employed to engrave them.

Like the maps of other travellers who have gone forth into the wilderness single-

---

* The Appendix contains a Notice of the objects of Natural History collected during the expedition, an Itinerary, and copious Notes on the present state of the Colony of the Cape of Good Hope.

handed, the author's has no pretensions to exactness, beyond that degree of it which was attainable with such instruments as he went provided with, and could use unassisted. It is, therefore, merely the result of a careful observation of his course during each day, and of the rate at which he moved, reckoning at three miles per hour the progress of a waggon drawn by bullocks, and at four, that of pack oxen. The daily tracks have been carefully examined by that able geographer, Mr. J. Arrowsmith, who has found but little discrepancy in the Author's distances in the interior, when compared with certain points laid down upon the coast by the marine surveyors; so that it is reasonable to presume that his work is as correct as his means would allow of. Indeed, he very much doubts the possibility of much more being achieved with a greater display of instruments, where travelling has to be performed over a surface so rugged, that no chronometer can be made to keep any thing like its rate, and where every other instrument is sure to be materially disqualified for use.

Should more information be required on particular points than is given in these pages, the reader is requested to bear in mind what are the difficulties attendant upon African travelling; the necessity there was for making personally every arrangement for the march, and for the night's halt; the trouble of procuring food, of dealing with the natives, of preventing drunkenness and quarrels among the people with whom the Author had to deal; and the constant exertion required to keep all in good humour. These duties so far demanded his time and care, as to forbid his doing more than is here set forth—" Quod potuit perfecit :" —he did what he could, and he now offers the description of a route of nearly four thousand miles, accomplished in little more than a twelvemonth, and all of which, to the north of the Kamies, or Lion Mountain, has never appeared in any former map of Southern Africa.

J. E. A.

London,
August, 1838.

CONTENTS

OF

THE FIRST VOLUME.

CHAPTER I.

CHAPTER II.

CHAPTER III.

CHAPTER IV.

CHAPTER V.

CHAPTER VI.

CHAPTER VII.

CHAPTER VIII.

CHAPTER IX.

CHAPTER X.

CHAPTER XI.

ILLUSTRATIONS.

VOL. I.

VOL. II.

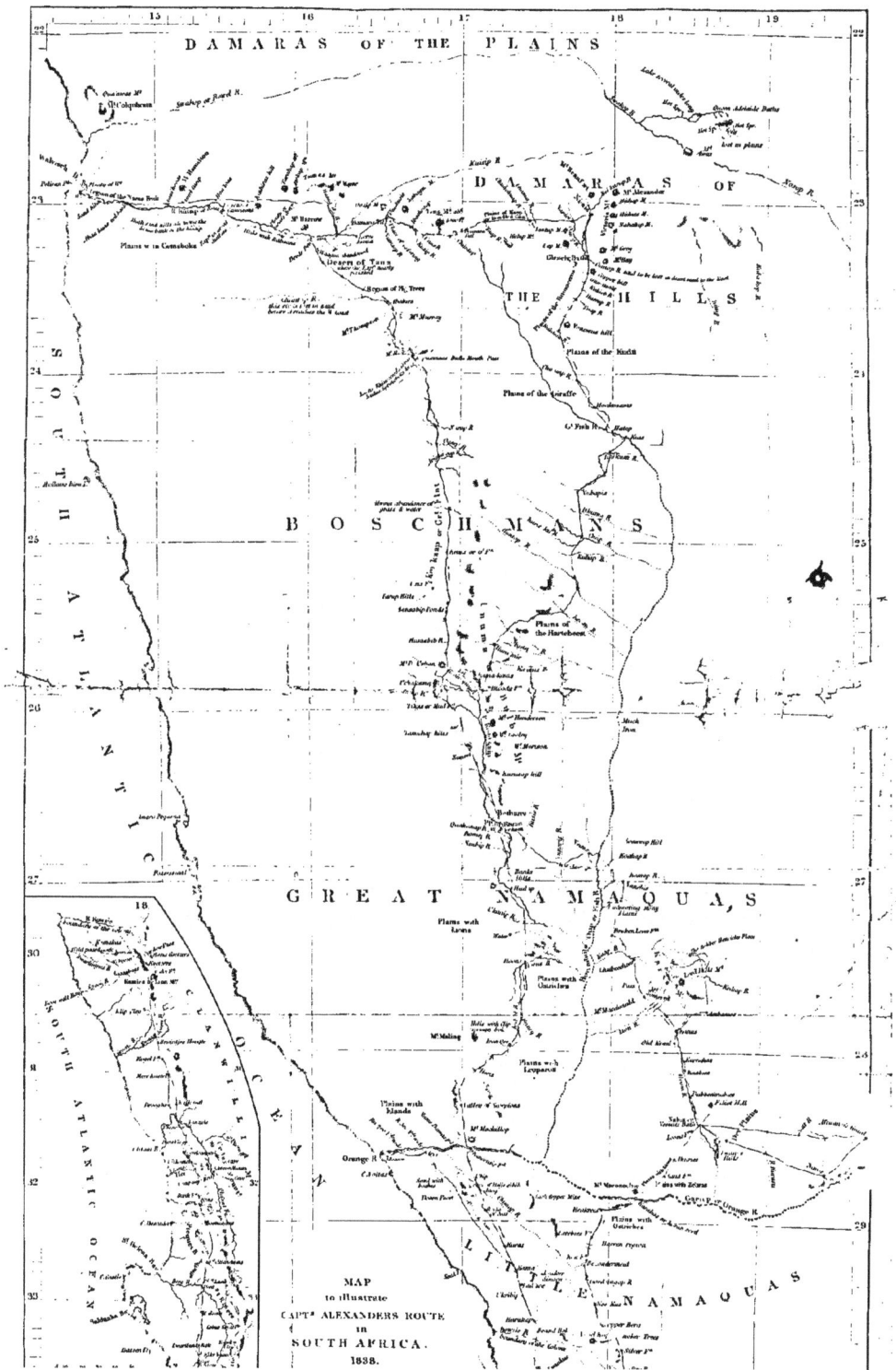

MAP
to illustrate
CAPTᴺ ALEXANDER'S ROUTE
in
SOUTH AFRICA.
1838.

EXPEDITION OF DISCOVERY,

ETC.

CHAPTER I.

Preparations for an Expedition of Discovery—My Attendants
—Leave Cape Town—Travelling Dress—Beautiful appearance of the Country—Difficulties and Accidents—
Different Modes of Travelling—Inhospitality at Malms-
bury—Field-cornet De Toit—Losing one's way—The
ready Cook—Misfortunes again—Anecdote of a Dutch-
man's Consequence—The Steenbok—Our Practice in
Hunting—Life among the Boors—The Berg River—
Mynheer Vandermerwe—Picquet Berg—Lang Vley—
Uneasiness of the Farmers—Ride to Clanwilliam—Tem-
perature—The Cedar Mountains—Description of them
and of the Trees—Two Excursions to the Mountains—
Boschman Caves—Boors' Habits—Murders—Return to
Clanwilliam.

At Cape Town, I employed the month of
August, 1836, in making my preparations for an
expedition of discovery into the interior of South
Africa, calculating that I should be absent at
least twelve months from civilized society.

I bought an excellent waggon, complete, with a tilt, fore and after chests, side cases, water kegs, yokes, &c., for 60*l.*; and a span of handsome black and white oxen for 40*l.* I did not expect to be able to take the waggon far beyond the Orange river; but when I thought I should be obliged to leave it   hoped that I might then have a sufficient number of pack oxen to carry me on.   Accordingly, I had seven pack-saddles and fourteen black leather packing cases made, to convey the necessary stores for our further progress after leaving the waggon.   These cost the large sum of 90*l.*; for materials and work-manship are expensive at the Cape.

Besides the saddles and cases, which took up little room, the waggon was freighted with the following articles: First, for offence and defence, three double-barrelled guns, four fusees, and three rifles, complete, with bayonets and swords; three pair of pistols, three boarding pikes, seven rockets, one hundred pounds of canister gunpowder, and a proportionate quantity of lead and pewter; two hundred flints, bullet moulds, bags of shot, powder horns, shot belts, and six hundred rounds of ball cartridge, for a stand, in case of being attacked.

Next, for bartering with the natives for food, there were saws, hatchets, files, brass wire, knives, tinder-boxes, needles, beads, buttons, cotton shirts, shawls, handkerchiefs, red caps, and one hundred and fifty pounds of tobacco, which last procures assistance and food, when all else fails to move the natives of South Africa.

The private stores, consisted of axes, spades, pickaxes, nails, spare yokes, and tar for the waggon; some rice, biscuit, salt, tea, and coffee; a medicine chest, clothes, stationery, books of geography and natural history; musical instruments, as a violin, tambourine, Pan's pipes, &c. to keep my own people "alive," and the natives in good humour.

Lastly, my instruments, furnished by Government, were a pocket sextant, small azimuth compasses, barometer, thermometers, artificial horizons of coloured glass and quicksilver, an Admiralty chronometer by Arnold, and a small air pump for extracting poison.

I did not invite any one to be my companion

in the wilderness; for after an experience of
fifteen years of voyages and travels, I found that
one always moves with greater ease, and infinitely
more at liberty, without a companion; and that
one is more induced to associate with, and to
collect information from, the natives of the
countries one traverses, when wanting an associate on the road: though, to counterbalance
all this, there are times and seasons when the
soul requires communing with a friend, whose
ear may be ready to listen to one's complaints;
whose voice may cheer, and whose counsel may
direct.   For a medical adviser I have fortunately
had no occasion for many a long year.   A con-
sort I hoped to obtain if I returned alive and
well!

From all I learned at the Cape, it was not
advisable to traverse the interior with native
attendants only: they might combine, for evil,
against their master; they might prove indolent
and cowardly; an admixture of Europeans was
necessary; and my party, therefore, consisted of
the following men:—

1. Charles Taylor, an Englishman, in charge

of the stores, and skilled in preserving objects of natural history.

2. Robert Repp, an Englishman, in charge of the cattle.

3. John Elliot, an Irishman, private, 27th Enniskillen regiment, in charge of the arms.

4. Antonio J. Perreira, my Portuguese servant.

5. Magasee, a Bengalee, who had been a slave in the Cape colony, and who had also lived eighteen years with the Caffers.

6. Henrick, the waggon driver, a powerful young man of the mixed South African race, or a " Bastaard," as the local term is.

7. Wilhelm, the waggon leader, also a stout Bastaard.

My people were bound to me for a year; and the highest wages were 3l. each per month, with food and clothes. It is not always easy, on any terms, to induce men to embark in an undertaking such as I now contemplated. I found that I had not a man more than I positively required.

Henrick's wife, Metjie, entreated to be allowed

to accompany him to the limits of the colony, which she was allowed to do, proving useful as a cook and washerwoman.

My preparations being completed after a month's hard work, and the expenditure of more than the government allowance (300*l.*) for the outfit of the expedition at the Cape, on the evening of the 8th of September, the waggon rolled quietly through the clean and tree-lined streets of Cape Town, passed between the mighty wall of Table Mountain and the castle on the shores of Table Bay; and as it passed the fortress, a gentle shower of rain fell, and the evening gun fired a parting salute.

The waggon waited for me at Laubser's farm, beyond the Salt river; and on the 10th of September I left Government House, probably for the last time, and a chief, Sir Benjamin D'Urban, to whom I had been long and warmly attached.

After breakfasting at the Royal Observatory, and partaking of a stirrup cup, I was honourably escorted thence for some miles by Sir John Herschel, Major Michell, K. H., the Surveyor General (now my father-in-law,) Dr. Murray,

principal medical officer, Mr. Maclear, astrono-
mer royal, Mr. George Thompson, the South
African traveller, and Mr. J. Wingate, a school-
fellow and esteemed friend.

I parted from these gentlemen (for all of whom
I entertained a high regard) with slight hope of
ever seeing any of them again; but resolved
to penetrate as far as my strength and means
permitted me to do, anticipating many strange
adventures; trusting to be of some service to
my country and to the aborigines of Africa,
and relying with humble confidence on a Providence by whom I had been hitherto mercifully
and strangely preserved.

I left the sand flats and ridges of the bay, and
with my half wild follower, Magasee, trotted over
grassy plains, passed some cultivation, and joined
the waggon at its outspan.

I found my people in good spirits, and pro-
fessing to be rejoiced to have escaped from the
confinement and dissipation of Cape Town;
than which few places are more destructive to
the lower classes, from the cheapness of wine
and brandy. The men were dressed in blue

smock frocks, over a cotton shirt, and leather trowsers; field shoes, or a sort of buskin, of untanned leather, and grey broad-brimmed and round-crowned hats, which, when we got farther on, were surmounted with ostrich feathers, as a protection from the sun. This costume, with belts and arms, gave them rather a picturesque appearance.

I was myself equipped in a similar manner, dispensing with the luxuries of stock, gloves, and socks, which are unsuited for rough work.

We inspanned, crossed the Mosselbank river, and passed on beyond Captain Proctor's farm. There was nothing of the desert in the appearance of the country at this season of the year, nor of aridity or barrenness observable; the face of nature being covered with a broad carpet of dark green, on which were patches of the most brilliant wild flowers. Cultivation was confined to the immediate neighbourhood of the farmhouses. On our right, the snow-capped peaks of the Drakenstein mountains, rising two or three thousand feet above the plains, formed a most agreeable picture. The Cape lark rose near us, perpendicularly, on whirring wing, to the height of about thirty feet, gently descending with a prolonged whistle; and all around us wore a delightful aspect of light and liberty.

But we were soon recalled from the enjoyment of the scene to very disagreeable duties. The roads were deep with mud, from recent rains, and the waggon was overloaded; as in setting out it was difficult to calculate what the oxen could easily draw; we consequently plunged into holes, from which neither shouting nor the whip—of both which the attendants were by no means sparing—could avail. By digging out the mud from before the wheels, tramping down bushes, unloading a portion of the stores, and other similar devices, we managed to surmount one difficulty, only to get into another; always looking, however, to a "good ending;" until, at last, we stuck fast up to the naves in a "slough of despond," from which there was apparently no hope of extrication.

It is the custom to sing the praises of the Cape waggons, their strength, their great length and pliability, preventing their upsetting;

their forming a convenient field house, &c.; but I confess I am no friend to the slow three-mile-an-hour progress of this conveyance, and its " stick-in-the-mud" propensities, though it probably suits phlegmatic temperaments. I had, however, no choice: there was no other means of transport to be got at the Cape; no pack oxen; horses would not probably live far beyond their accustomed pastures, and no one has yet had the enterprise to send from the Cape to Bombay for camels, to try if they will live to be useful in the plains of Southern Africa.

I went on ahead, and found mud holes every where. I saw that I must call on the field-cornets for assistance, and make use of my government order and passport, or else return to Cape Town to buy another waggon and a second span of oxen. I preferred the former course, and rode forward to the small village of Malmsbury, where I arrived at dusk; and proceeding to a house to which I had been recommended, I was furiously assailed by dogs and the shrill voice of an old woman, intimating that the master was not at home, and desiring me also to

"loop," or take myself off. From thence I went
to the field-cornet in the village, and he, making
an apology for the smallness of his house, directed me to another field-cornet's, two miles off.
Again "I took the road," and turning to Magasee, I told him to prepare for passing the night
under a bush, in our old Caffer style, for that I
saw no prospect of getting either assistance or
shelter that evening. But on arriving at Mynheer De Toit's, he immediately asked me to "off
saddle," and gave me the best his house afforded,
with a capital bed.

Next morning our obliging field-cornet sent
a messenger on with an order to a Mr. Dekok
to assist us, and we followed after; but we were
again unfortunate; it rained heavily—we lost
our way, went a round of twenty miles, and
arrived at a Michael Dekok's, no great distance
from Cape Town. Here we broke bread, and
were set right; found the proper Mr. Dekok's,
and transferring a large portion of the goods from
the expedition waggon to another, by means of
an extra whip the bullocks did their duty, and
we tarried for the night at Mr. Jan Dekok's.

" Sir," said one of my men, "there's the readiest black cook here that ever I seed; she heard that the Governor was coming, and she has been ever since in a terrible bustle, running here and there with a roaring child on her back. I asked her for a light for my pipe, and she whipt off a piece of her shift in a moment, rubbed it up, greased it, and struck a light with it. She then put down some soup afore me, tore off another rag, cleaned a spoon and the table with it, and then stuck it into her sleeve for another time. She's never at a loss for any-thing, sir !"

On the 15th we got on well with an extra horse waggon; passed through Malmsbury, and by Zwartland Kirk, the centre of the " Brod kammer," or bread-room of the colony, the land here being very fertile, and producing corn in abundance. We thought now we should have no more stoppages, when again both the horse and ox waggons floundered into a deep quick-sand, and there was again delay and hard work, roadmaking and unloading, but we again pro-gressed, when in crossing a stream the fore bolt of the expedition waggon broke, and we were brought up again.

I thought, as the Persians have it, that "we had not set the lucky foot foremost," when we commenced the journey. However, as long as we were not positively compelled to return to the Cape, I did not much care, but galloped to the next farmer's, Martinus Smits, bought a bolt from him for three rix-dollars (four shillings and sixpence), and returning to the waggon, shipped it, with the assistance of some Boors, who in broad-brimmed hats, cloth jackets, and leather trowsers, and provided with long guns and immense powder-horns, were with horse and hound in chase of steenboks.

A characteristic anecdote may be here told of one of these Dutchmen. In my absence he was approaching the waggon, and accosting one of my coloured attendants, he asked him if he had seen a flock of sheep pass that way?

"What say ye?" replied my man, without the affix of mynheer or baas, (master).

"Ye, to me!" cried the Boor. "I'll teach you better manners; I'll get you punished.

Come along to your master with me immediately."

The indignant Dutchman approached the waggon, and addressing himself to one of my white
men, who was busy inside packing, he began to
lodge his complaint, on which the other, without looking at him, and not understanding a
word he said, coolly told him to "go to h—l!"
which still more disconcerted the ill-used boor.

The steenbok (or stone buck), of which the
boors were now usually in pursuit, is a small
but very graceful antelope, three feet and a
half in length from the muzzle to the tip of the
minute tail, about one foot eight inches high,
with long and slight legs, of a reddish fawn
colour, with a beautiful gloss. The female is
without horns; those of the male are much
shorter than the ears, and are straight and very
sharp. My friend, Mr. W. Ogilby, F.Z.S., was
the first to notice the absence of spurious hoofs
in this elegant antelope.

The steenbok (singly and in pairs) frequents
the broad plains on which are scattered rocks
and low bushes. It seemed to me always to be

a very foolish animal; for it would run a short distance, and then turn round and gaze at the sportsman, who openly approached it. Yet it was painful to see the poor creature, after receiving its death wound, circle round a bush with drooping head, before it hid itself to die. Its flesh is very delicate.

Here I beg to remark that during the whole journey nothing was killed wantonly, or that we did not positively require either as an article of food, or as a rare object of natural history. I never could, and I trust I never shall, reconcile myself to the notion which some sportsmen entertain, that it is manly to destroy as many animals as one can: this thirst for blood is discreditable. The exploit of which I know some to boast, viz., killing four elephants in one day, or the same number of hippopotami, with the same gun, for mere sport, is surely not praiseworthy.

I bought meat and bread for my people at M. Smits, which they ate at the waggon, whilst I was invited into the house to partake of the country fare. I have a great aversion to greasy food; and to please the Boors, and thereby get more smoothly through the colony, I forced myself to eat of their dishes, swimming as they were in sheep's tail fat. But being no Fuseli, may I never again have the horrid dreams I had after this fare, and in disagreeably soft beds!

The Boors thought I was going to look out new locations for them beyond the boundary, and they wondered at my temerity in proposing to cross "de limite." I had a number of set questions for them, by which conversation was kept up for a considerable time: such as interrogatories regarding their children—the smut in corn—the horse sickness, and the advantage of substituting wooled sheep for the bare-backed large-tailed breed.

On the 16th we reached the Berg river, filled from "bank to brae," and flowing through a flat and fertile country. The strength of the stream prevented the punt crossing, we saw twenty waggons collected on the other side, and we were compelled to halt for the subsiding of the waters.

Mynheer Vandermerwe, of the punt, was a strapping fellow, with red hair, a loud voice, bearing a walking-stick in his hand as long as himself; that is, six feet four: though a field-cornet, he could not read my passport, and had conveniently lost his " bril," or spectacles. After our supper of fatted mutton and stewed peaches, a tub of water was, as usual, passed round the company, in which, beginning at the Baas, and without changing the water, the feet of half-a-score people were washed by a slave girl.

Crossing the Berg river, we were under the Piquet hill on the 19th. This is a long primitive eminence, with heavy sand at its base, through which we slowly moved. The Piquet berg is supposed to be a good locality for the growth of coffee. There is an excellent garden here at Bomzagers farm, which is pleasantly situated under the picturesque hill, and looking towards the east: quince hedges enclose fruit and vegetable ground, and we purchased a hundred oranges for one shilling from the proprietor and his vrouw, who were sitting on their stoep or terrace, in front of the house, with its ogie gables,

and enjoying the cool and quiet of the evening—
" procul negotiis."

Losing our oxen for some hours, and with other
trials of temper and patience, becoming also
rougher and more bush-like daily, " oblivious of
silver forks and kid gloves," we journeyed on-
ward, occasionally leaving the road to follow
steenboks, korhaan or bustard, and enjoying our
gipsy outspan in the middle of the day by the
side of bushes and near water, where we could
conveniently swing our pot on its own " three
legs," and eat our frugal repast off the grass.
The Turkish poets wisely recommend us to enjoy the passing hour to its fullest extent, and
thus one sings of the spring.

> " Tulip, rose, anemone,
>   Blossom in the sunbeam free,
>   Watered by refreshing showers ;
>   Yet we find their vernal treasures
>   Vanish like life's fondest pleasures ;
>   Age and care fill up the hours :
>   Then seize, oh ! seize while rapture flows
>   The season of the vernal rose !"

On the 21st we were at Lang Vley (Long
Lake), Piet Vanzyl's ; thence I sent the waggon,
a few hours further, to Uitkomst (Coming out),

H. Vanzyl's, there to remain for a few days till I had visited the village of Clanwilliam and the Cedar Mountains.

At this time, in the colony generally, there was a very restless spirit abroad. The farmers were dissatisfied with the government on account of the sudden emancipation of their slaves, and on account of the losses the Eastern farmers had sustained by the Caffers. They were also generally filled with exaggerated notions of the beauty and fertility of the country to the N.E. of the boundary, and West and N.W. of Natal.

" We read in the papers," said a Dutch farmer to me, " that in Europe we are considered as tigers, and that we destroy the coloured people without mercy: look round and say if you see any thing of this. We are vexed and annoyed at the opinion which is entertained of us; and no allowance is made for us, for being reluctant to lose two slaves out of three, for we are only paid for one out of three; thus a farmer who gave a few years ago 800*l.* for a few slaves to cultivate his ground, now receives only 300*l.* for them. Our countrymen too on the eastern frontier have been ruined by the Caffers, and have not recovered their property.   We hear of the great fertility of the land beyond the north-eastern limit, and we wish to try and find out a new country for ourselves."

I tried to sooth the farmers as well as I could, said that there was a hope that free labourers would be more profitable to them on wages than the slaves had been without—that if they abandoned the " oud plaats," they abandoned the comforts and security of civilized life, would expose their wives and children nightly to the attack of savages, and must continually sleep with their fire-arms in their hands.

I got up early on the 22nd, and rode over lonely hills and through silent vallies to the Olifant (Elephant) river, which ran deep and full through a sandy plain.   I met Mr. Ryneveld, the assistant civil commissioner or chief magistrate of the sub-district of Clanwilliam, on the western side of the river, and with him I forded it, and shortly afterwards arrived at the village of Clanwilliam, where I took up my residence with Mr. Ryneveld.

Clanwilliam is a village of neat houses, one story high, and arranged in two streets. The number of inhabitants is about two hundred. There is a church here, and a few shops; and the better sort of houses are provided with good gardens; but situated as Clanwilliam is, in a basin-shaped valley on sand, and surrounded with hills (whilst the grand summits of the Cedar Mountains tower in the distant S.E.), it is perhaps the hottest place in South Africa.

Register of the thermometer in the hot months.

	9 A. M.	Wind.	Noon.	Wind.	3 P. M.	Wind.	Observations.
1830 Dec. 5...	82	E.	86	W.	86	N.W.	Clear.
— 15...	76	N.E.	91	W.	90	W.	Clear, with strong wind.
— 25...	76	S.W.	78	S.W.	84	W.	Calm and clear.
1831 Jan. 10...	84	N.W.	96	S.W.	90	W.	Cloudy.
— 16...	88	E.	106	E.	100	N.E.	Clear.
— 22...	92	E.	110	N.E.	105	N.E.	Do.
Feb. 5...	94	N.E.	112	E.	110	E.	Do.
— 15...	79	E.	92	N.E.	88	S.E.	Do.
— 24...	83	S.W.	95	S.E.	100	S.E.	Thunder storm— with heavy rain.

The heat, it will be observed from the above short table, must be very distressing at Clanwilliam in December, January, and February. Those of the white inhabitants who can afford it,

live during these months on farms, or among the
Cedar Mountains, some distance from Clanwilliam. The nights in the hot season at Clanwilliam are often more stifling than the days,
sleep is denied, and bottles sometimes burst in
the rooms with a loud explosion, from the heat.

I was not long in Clanwilliam before I mounted
and rode with Mr. Ryneveld towards the beautiful range of the Cedar Mountains.

This primary range of mountains, running
north-east and south-west, consists on the upper
parts of ash-coloured quartzose sandstone, whilst
the lower parts contain a variety of mineral and
fossil substances, particularly marine petrefac-
tions, as shells and fish; also black, red, and
striped jasper, hornblende, crystalized carbonate
of lime; garnets imbedded in argillaceous schist;
opaque and crystalized quartz, beautiful varieties
of agate, and iron ores of different kinds, and in
such quantity that all the water, of which there
is abundance about the Cedar Mountains, is
more or less chalybeate.

The valleys are rich in decomposed vegetable
matter, which forms a fine dark-coloured mould,

which is exceedingly productive. The chief produce of these fertile spots consists of corn, tobacco, and fruit. A small quantity of wine is made here; and altogether, the Cedar Mountains, though not extensive, are perhaps the most interesting range to the geologist in South Africa.

The height of some of the highest peaks, as calculated by the Baron Von Wurmb, (who visited this part of the colony a few years ago to found mission stations for the Rhenish Society), is as follows:—Rondeberg 2,990 feet, Groenberg 4,860, and Sneeuwberg 5,000.

I was particularly interested about the cedars, "the glories of Lebanon," which formerly used to clothe, and do still in part adorn, the higher glens and ravines of these mountains. These noble and imperishable trees are always attractive to the readers of scripture; for with them was the magnificent Temple of the wise Solomon built; but alas! for the cedars of these mountains, they are fast disappearing under the destroying hand of man. Annually the Dutch farmers, the Hottentots, and Bastaards, living

about the mountains, burn the grass to improve
the pasture.   Many fine old trees are thus sacrificed, and hundreds of young ones.   No one
takes charge of the trees, nor is a word spoken to
save them.   Their existence is almost unknown
at the Cape, where the wood would fetch a high
price; and even in the district where they grow,
two skillings or sixpence are paid for a foot of
planking, and ten dollars or fifteen shillings for
a beam.   To give an idea of the size of some of
these trees, a wide-spreading father of the forest
was cut down in 1836 which was thirty-six feet
in girth; and out of whose giant arms 1,000
feet of planking were sawed. ·

I took care to represent to the government at
the Cape the expediency of superintending the
cedar trees; of preventing their being mercilessly destroyed as at present; and I advised
that seed should be continually collected, and
sown in proper situations on the mountains.

I made two excursions to the Cedar Mountains: the first was in company with Mr.
Ryneveld.   We rode through valleys of the most
grotesque forms of rocks, standing bare after

their softer beds had been long wased away from about them. Some were arranged like a great arch; others overhung the ground below at a sharp angle; some were balanced on a single point; and others formed an open klip, (or stone) house. A narrow gorge is called the " De Smits Winkel" (or shop), from the ringing noise the rocks give out in passing through it.

These huge masses of old sandstone were stained with oxide of iron. There was grass and shrubby plants among the rocks; and here there were broad green leaves encircling the purple flowers of the protea mellifera, or sugar-bush of the Cape.

Leopards inhabit the mountains, also bucks of various kinds, wild boars and baboons: there are also many partridges and pheasants, and insects in considerable variety, in the antipodial spring months of August, September, and October.

There is a Rhenish mission station called Wopertal in a valley of the Cedar Mountains, where the climate is very pleasant, even in the hottest months. There is good ground, and plenty of wood and water at the station; and the small Hottentot community, under Mr. Lepold, are employed in gardening, carpentry, and shoe-making, and are industrious and thriving.

Mr. Ryneveld took me to a retired and beautiful spot in the mountains, called Reebok Valley. Here a little old farm-house was shaded by oaks; the garden contained rows of pear and peach trees, and vines were set in good soil; poplars grew beside a water-course. Here we partook of thick milk, with sugar and bread, and enjoyed the seclusion of Reebok Valley.

My next excursion to the Cedar Mountains was with an agreeable and worthy man, Mr. Wynzel, land-surveyor, to whom I subsequently sent olive plants and grass-seeds, that the emblem of peace might be introduced in the western districts of the colony, (as I had formerly endeavoured to do in the eastern province); and that the pastures also might be improved.

We put up our horses at Nieuwouds farm, and then, in company with Tim, an old English dragoon, proceeded to examine some Boschman caves, three or four hundred feet above the level of the plain. In clambering among the rocks we saw in a hole small long-eared bats, also a yellow-snake, and the traces of dassees, or rock rabbits, whilst an eagle or falcon, which preys upon them, sailed majestically up the valley, casting its keen glance towards us.

The Boschman caves consisted of overhanging rocks, were thirty or forty feet across the entrance, and did not extend above twenty or thirty feet into the mountain from the entrance. The roofs were black with smoke. The caves were deserving of notice from the paintings upon them. In one I saw, not far from Nieuwouds farm, there was a flock of sheep with their lambs represented in red ochre, the outlines of which were surprisingly accurate; whilst higher up is another cave, in which Boschmans are seen combating with javelin, bow and arrows; these traces of a rude people, who have long since disappeared from this locality, are very interesting.

At the farm-house a clever-looking old woman, eighty-one years of age, presented us with cakes made with flour and mountain honey, which were very palateable. Beyond the Cedar Mountains, in the Bokkeveld, there was, in 1836, a man 110 years old; but instances of longevity beyond seventy years are rare in the colony. The people are perhaps like African oaks—they soon come to maturity and soon decay, having no winter to check their growth; though I often thought that people here ought almost always to see one hundred years, living as they do in so fine a climate, where there are no swamps, or malaria, or fever. But men and women live grossly; that horrid sheep's tail fat clogs the wheels of the machine. Besides, the men take, in general, little active exercise, and the women less, the former leaving any hard work to slaves, and only taking off their coats occasionally to drive a waggon, whilst the vrows sit in a corner preparing " tea-water," and get fat from inactivity, and often irritable from the carelessness of their female slaves. What a cursed taint slavery every where carries with it !

What made some noise in the sub-district of Clanwilliam at this time was the murder of a coloured woman and two children at a farm-house, in the absence of the master, by two runaway slaves, who were desirous of getting arms to raise a gang of robbers; but the murderers were eventually secured and executed.

Crime is here more frequent than it would otherwise be, through the disinclination of farmers to prosecute, as they have perhaps to ride a distance of fifty or sixty hours, and the same distance back on bringing an offender to justice.

I returned to Clanwilliam, where I experienced every mark of hospitality and kindness from the family of Mr. Ryneveld.

CHAPTER II.

Account of the Sub-district of Clanwilliam — Its Extent
and Capabilities — Objection of the Dutch to Merino
Sheep — Irish Settlers — Culture of the Vine — Number
of Farms and Population — Wages — Villages — Roads
—An Extended Boundary proposed — Dutch Grants —
A Military Post or Police Station recommended — Clanwilliam ought to be a district by itself — The reasons for
this — Two great wants for the Body and Soul — Resume
our Journey — Dutch Hospitality — Uitkomst — Heere
Logement — F. Vailant — Elephants — Ebenezer Mission Station — Inspect the Mouth of the Olifant River
—Its Inundations — Movable Field Table — Kalk Gat —
South African Morning — Paddegat — Namaqua Par-
tridges — The Grease-pot — The glorious liberty of the
Bush and of the Road — Battle of the Bees — Zwaart-
doorn River — Nieuwouds Field Station — Leopards and
Boschmans — An Accident — The Groene River — At-
tention to Travellers — Snakes — Sunrise among the
Mountains—Little Namaquas—Arrive at Lily Fountain.

THE sub-district of Clanwilliam is from
north to south about 300 miles in extent,
and from east to west 250. The fertility of
all countries depends on rain : so here, after
very abundant rains, the Olifant river, (the
principal one in the sub-district), overflows its

banks, and then the produce is very considerable. The wards (or the divisions of the country under the charge of a field-cornet) at the mouth of the Olifant river, the Onder Bokkeveld, Bedouw, and Hantam, are considered the most fertile portions of Clanwilliam, whereas the Hardeveld is the most barren; the other wards are of a middling description.

In the wards Bedouw, Onder Bokkeveld, Hantam, and also in the Onder Roggeveld, much corn might be raised; but as the roads are sandy and heavy, and no navigable rivers, the land is used chiefly for grazing cattle. These wards also, as well as the wards of Olifant river, Berg, Lang Vley, and Clanwilliam, are well adapted for the breeding of Merino sheep.

The Dutch farmers have an objection to Merino sheep, by which stock, however, some English settlers in the Eastern province of the Cape, and many in Australia, are making fortunes. The Dutch say that they cannot be troubled with washing the wool; and that these sheep have not the fine fat tails of their own;

but if I succeed (which I have several times
tried to do) in introducing olive oil into the
colony from trees grown on the soil, this will
supply the place of the sheep's tail fat.   In
Clanwilliam, as in other parts of the colony,
there is abundance of room for English settlers,
and Clanwilliam is far removed from the verge
of Caffer inroads.   Some years ago a considerable party of Irish settlers was sent to Clanwilliam;   but they were under bad management; and the land allotted to them was barren,
and in a very hot situation.   It is, therefore, not
to be wondered at if they turned out ill and
dispersed.

The culture of the vine is at present only
practised in the wards Berg and Lang Vley,
Upper Olifant River, Clanwilliam, Bedouw, and
on a few farms in the wards at the mouth of
the Olifant river; but it is considered that it
could be cultivated to any extent, and with great
success, particularly about the mountain ranges,
where there is always abundance of water.   The
colony generally might be made πολυτα φυλον, or
" grape-abounding."

At present 370 places or farms are occupied in Clanwilliam which pay taxes; but there are many more which pay none for want of being surveyed. The Surveyor-General's department is not sufficiently large for so extensive a colony, and particularly since the old (pretended) surveys are so faulty. Farms are now found to overlap each other on the map in the most extraordinary manner. With an importation of emigrants, the strength of the department ought to me much increased.

The population of Clanwilliam is as follows:

Whites.		Hottentots.		Apprentices.		Total.
Males.	Females.	Males.	Females.	Males.	Females.	
1189	1115	2566	2445	575	570	8460

But it is well known that this district could support a much greater population, which at present is very trifling compared to its great extent.

The payment of the free coloured farm servants is from six shillings per month to nine shillings. They also get clothes and tobacco periodically; and meat, meal, and vegetables daily. With the exception of a few, their character is not very good.

I inquired for sites for villages, and was told that a village could be formed in the wards Berg and Lang Vley, a short distance from Lambert's Bay, and also at the mouth of the Olifant river. Villages would increase the comforts of the farmers, and extend knowledge by the establishment of schools. Schoolmasters are few, and generally very indifferent in the Cape colony.

Great improvements with respect to roads have already been made in Clanwilliam within the last few years; and much more might be done (without much expense to government) to render travelling from the remote parts of the district to Cape Town easier than it is at present. Punts and boats ought also to be established on the Olifant river. Thus for want of a boat at Clanwilliam last year the post was twice carried away: the first time he escaped, with the loss of his mail-bags; but the second time he was drowned.

Some propose to extend the boundary of the colony to the Orange river, for the following reasons. There are a number of farmers living beyond the Kowsie, or Buffalo river the limit of Clanwilliam, who have paid taxes to government for the last thirty years; these or their predecessors obtained their farms in the following strange manner. In the Dutch time they applied for places *beyond the Olifant river*. Now, in these days, the information of the Cape authorities was very limited regarding the geography of the colony, and matters were conducted in so careless a manner, that the farms in question were granted, and it turns out that, not only are they beyond the Olifant river, but beyond the boundary also, which the applicants well knew when they applied for them. These farmers continue to pay taxes, that they might have a claim on colonial protection.

Secondly, it is proposed to extend the boundary to the Orange river, because the present limit is badly defined, and in some places it is impossible to say where Clanwilliam ends.

Thirdly, it is proposed to extend the boundary, because in seasons of drought, colonists, white and coloured, itinerate beyond it to feed and water their cattle.

Yet in all this the rights of the aborigines are overlooked—the Bastaards and Namaquas; but if these people are willing to be received into the colony, and think that they would thus be better protected against robbery and oppression among themselves; then there would be no objection to the extension of the limits.

A small military post, or police station, manned by steady Bastaards or Namaquas, might advantageously be placed on the Kamiesberg, or else somewhere near the frontier, to look after the Boschmans, or other troublesome neighbours.

One thing seems to be particularly required; viz., the erection of Clanwilliam into a district, instead of being a dependency of Worcester, as at present. It is impossible for the civil commissioner of the latter to become acquainted with the wants of the inhabitants of such an extensive sub-district as Clanwilliam, except through the assistant civil commissioner. The distance, then, between Worcester and Clanwilliam being one hundred and sixty miles, great delay unavoidably and constantly takes place in the despatch of business; and in the event of hostile inroads, from the barbarous tribes beyond the boundary, very alarming results might take place before information could be given at Worcester, and the necessary orders returned from thence.

Again, much unnecessary trouble is occasioned to the government servants at Clanwilliam by the circuitous route by which all accounts have to pass before they can be finally settled. The assistant civil commissioner must first certify to their correctness, then transmit them to the civil commissioner at Worcester; by whom they are sent to the auditor general, when drafts of payment are forwarded to Worcester, and thence to Clanwilliam; so that when those who have claims on government are ultimately paid, this unreasonable delay causes great dissatisfaction to the parties.

Then, in case of memorials for redress of grievances, great delay takes place: thus, a memorial is sent to Worcester—the civil commissioner being in doubt as to the correctness

of the statement, it is sent to the assistant civil
commissioner for his opinion, who then has to
send it back; when the civil commissioner frames
his report to government. Thus many weeks
elapse before the memorialist can ascertain the
fate of his petition.

But it is time to leave Clanwilliam village,
and to go on with our expedition.

Before I proceed, however, I must notice two
important points. First, there was not, when I
was at Clanwilliam, a single medical attendant
(government or private) from Lily Fountain,
near the northern border, to Cape Town; not
a person who could set a limb, or heal a wound,
in a distance of four hundred miles! The care
of the body being thus little attended to, it is
not to be expected that the soul was much looked
after.

Secondly, children of three and four years of
age were not baptized in many parts of the district. The parents were most anxious to have
their children baptized, and to hear occasionally
the Gospel of peace themselves; and often offered to send, and did send, waggons to bring a

clergyman to them, but no one, for four or five years, had made a progress through the district; whilst in the village of Clanwilliam itself, I noticed much vice and debauchery among the coloured and *uncared for* part of the population. All this loudly calls for reform.

My people were very well attended to by that most hospitable Dutchman and worthy farmer, Henrick Vanzyl, at Uitkomst. Meat, bread, and milk, they got in plenty, and for which no remuneration was asked or expected; and I fear that my small present to the kind-hearted vrouw inadequately repaid this family for all the trouble they had with us. May their prosperity greatly increase!

We inspanned the waggon, and, refreshed with showers of rain (which here did us no damage, as the soil was sandy and unlike the rich mud of Zwartz land), we journeyed onwards through flower-decked coppices, with a range of party-coloured mountains on our right. We arrived at Heere-logement (or Gentleman's Lodging), where was a pool of water, under a hill, some distance up which is a large and open cave, or klip-huis.

A small tree grows out of the fissure of the rock above, and partly overshadows the floor of the cave; whilst, on the north side, is carved the names of travellers and hunters, from the year 1712 to recent periods. Among others, conspicuous, is that of the renowned

F. VAILANT, 1783.

Looking from the cave in a westerly direction, the eye ranges over a wide extent of plain, on which bushes are scattered.

On the first of October we were on the banks of the Olifant river, which ran full and clear between steep banks which were lined with mimosas and willows. Elephants have long since disappeared from this locality; the only traces of them being the name of the river and the rude figure of one I had seen carved in the klip-huis at Heere-logement.

I turned to the left and went down the river twenty-four miles to Ebenezer, a Rhenish mission station. Here I found a small thatched chapel, dwelling-house, and school, and the rude huts of some Hottentots near the river. The missionaries, Messrs. Knap, Terlinden, and Hande, also Mr. Lepold of Wopertal (among

the Cedar Mountains) gave me a kind reception.
I stayed with them two days.

The site of the institution of Ebenezer was
formerly called Doorn Kraal (thorn village or
pen) and was an old location of a tribe of Hottentots under a captain of the name of Louis.
He and his people professed a desire to have
missionaries with them; the land, amounting to
11,180 acres, was surveyed by government for
them, and the German Baron, Von Wurmb,
assisted with some missionaries, founded this
institution in 1833, and that of Wopertal.
At Ebenezer, there are at present only one
hundred and eighty Hottentots on the books,
but if the missionaries succeed in leading out the
waters of the Olifant river over the land, and
obtain additional grazing ground on the opposite side of the river, in which I assisted them
as much as I could, there is little doubt that the
institution will thrive.

I got my goods ferried over the river in a
boat; and then placing a large bundle of reeds
and two empty casks in the waggon, whilst the
side cases steadied it, it was hauled over by a

rope, and when it grounded the oxen were inspanned, and it was dragged up the north bank.
This transit being completed, I left the waggon
to go on to the next outspan place; and then
proceeded to inspect the mouth of the river with
two of the missionaries.

Avoiding the sinuosities of the river, we rode
for two hours at a very rapid rate; and then
from an eminence saw below us the river di-
viding itself into two branches, enclosing an
island, and then emptying itself by one mouth
into the South Atlantic.

There are eight, ten, and twelve feet water
inside the mouth, which is obstructed by rocks
on which the sea beat angrily in white foam; on
the bar there are only six feet water. Flocks of
gulls, sea swallows, and flamingoes desported
themselves on the still waters of the river; and
it was evident that if one of the branches could
be shut up, and the whole body of the water
directed through a cut, which might be made
through two hundred fathoms of sandstone (to
the north of the mouth) that an opening might
be made into a small and secure bay, and thus

the river be made available for the shipment of produce.

In the Bokkeveld, muids* of corn lie unsold for years owing to the distance from the Cape market. In the Bokkeveld wheat may be bought generally for four rix-dollars, or six shillings the muid, whilst it sells in the Cape generally at twelve or thirteen. If the Olifant river mouth could be opened, much more corn would be raised in the Bokkeveld, &c. the farms there would be materially benefited, and the price of corn would be much lower at Cape Town.

The Olifant river rising behind Tulbagh, and flowing north and then west, is fed by several constant mountain streams. Sixty miles from its source it is joined by the Jan Dissel river; thirty miles below that by the Doorn river; after another thirty miles, by the Hol river, from which the mouth is distant twenty-five miles. In the basin inside the mouth " harder" and "springer" fish are to be caught in plenty.

The valley of the Olifant river is generally flat; therefore with much rain the river, as I

---

* A muid is 200lbs.

said before, overflows its banks. Rich karroo mud is carried down with the flood, which, upon the waters subsiding, is lightly sown with corn, and the produce is then more than one hundred fold.

It is to be regretted, however, that though there is always plenty of water to lead out over the lands, that the inundations of this South African Nile are not regular, there is no flood for three or four years sometimes; the average of considerable inundations is once in four or five. When these do occur, however, the vegetation is so rank that it is impossible to sow seed in handsful, but only by sprinkling it between three fingers here and there, and then the stalks and heads of corn appear in due time like bushes, thick and heavy.

We rode down to a small mat hut near the mouth, beside which a waggon stood. I found there old Henrick Vanzyl, with a fine tall light-haired and blue-eyed youth of a son, splitting, salting, and drying fish. A Hottentot woman soon fried us a few, which we ate with relish; and then shoving off in a boat, we were rowed to the north bank (to have a better view of the mouth) by an old soldier of Napoleon's who had fought at Austerlitz, Jena, Dantzig, &c., had been a prisoner at Chatham, and was also for six years in the British 60th Rifles.

Talking of a Hottentot woman reminds me that there is in this district a rival of the Hottentot Venus, if she does not excel her in the quantity of " cebaceous deposit." Rewarded by a trifle of money or tobacco, she will good naturedly allow a cloth to be spread behind, and on which four plates may be laid, thus forming a peripatetic table !

I again turned up the river, and arriving opposite to De Toit's farm, I crossed over, and with a guide, rode on in the dark, to the outspan of the waggon at Kalk gat, or lime hole, and found the people cooking and enjoying themselves round a couple of fires, behind the bushes.

Next morning, at half-past one, our Dutch conductor having a stomach ache, and not being able to sleep, awoke us, and said we must now inspan to go to the next water. At this com-

fortless hour, then, we got the cattle together,
and were under weigh. I walked drowsily behind the waggon, shivering with cold. No one
was in a good humour—my people thought of
the comforts of a house, of regular sleep, and of
the inconveniences of travel. Presently the
moon rose and sailed for a time overhead; the
approaching dawn was announced by the long
boom of a beetle, repeated at intervals as we advanced; and when the first faint trace of sun-
light appeared, a single bird chirruped from
bushes, exhaling a perfume of aniseed; the
song of birds increased till the sun rose, when
we halted, and threw ourselves on the ground
to repose.

Our next stage was to Paddegat, or Frog
Hole, eleven hours, or thirty-three miles. We
had left houses behind us at the Olifant river,
and on entering Little Namaqua land we traversed a wild region of bushes and brackish
water. After drinking this for some time, good
water tastes very insipid. At the pools, flights
of Namaqua partridges rose noiselessly from the
stony ground, and in coveys of eight or ten,

winged their whirring flight in gyrations through the air. We shot many, and found them plump, but tough eating. These birds are improperly called patrijs, or partridge; for they are grouse with three toes; their colours are brown speckled with white; and the tail ends in a point.

I allowed of no more stomach aches, and we got up now at the more reasonable hour of four, and then inspanned and travelled till the sun was warm. It is not to be wondered at that the Dutch are occasionally annoyed with bowel complaints, from the gross manner in which they swallow grease of all kinds, pouring spoonsful of melted sheep's tail fat over their food, and heaping butter in lumps on their bread. A supply of butter I had bought at a farm house to last us for a week, disappeared at one sitting, before two young boors invited to partake our evening meal.

I now felt all " the glorious liberty of the bush and of the road." I could dress as I liked, could rise and lie down when it suited my pleasure, went fast or slow, sang aloud or kept silent, ate my food with an appetite of the keenest Savigny edge, and was gratified with the appearance of picturesque hills and broad and verdant plains, was cheered with the sight and sound of birds and insects, while lizards of various colours, with yellow or green scales, tipped with red and gold, continually hurried across our path, or an occasional snake would glide among the stones and bushes, with its striped or spotted skin.

On the road, lightheartedness is indispensable.

> " He who would happy live to-day,
> Must laugh the present ills away,
>   Nor think of woes to come;
> For come they will, or soon or late,
> Since mixed at best is man's estate,
>   By Heaven's eternal doom."

My people brought me a piece of honeycomb, and reported that there was a large supply under a rock. We proceeded to the spot to smoke out the bees with burning bushes; but Antonio imprudently poking the nest with a stick, the swarm flew out upon us in a rage. I called out to the people to keep down their hands and run. We scrambled down the rocks, hotly pursued by our foes, whilst all those who tried to knock off

the bees were terribly stung; and Mr. Antonio
and others were forced to plunge their heads
into a pool of water to get rid of their tormentors. Thus ended the " Battle of the Bees;"
which simple affair was afterwards magnified, at
the Cape, into the imprudence of one of the
party occasioning an attack on us by the aborigines of the country—wounds and defeat!

Travelling through broad and narrow vallies,
and over green slopes bounded by hills, we reached
the Zwart Doorn (Black Thorn) river, in whose
sandy bed grew numerous mimosas. In a sheltered nook among the hills were two large circular huts, or wigwams, and three smaller ones,
composed of bent boughs neatly covered with
yellow rush mats; by them stood a couple of
waggons; there were also circular kraals of
bushes, for cattle. This was the field residence
of Mynheer Nieuwoud.

The part of the colony we were now traversing is drier than others, the farmers, therefore, have each at least two places of perhaps
three or four thousand morgen (six or eight
thousand acres) each, and to save the pasture

about their houses for summer, they are in the
fields, at a distance from their homes, with their
cattle, during the months of July, August, September, and October, and move about from one
pasture to another, in the Tartar fashion.

We outspanned near the field-cornet's, who
came out, and saluting me, invited me into his
rush tent.   Nieuwoud was a very burly man, in
a broad-brimmed hat, blue jacket, and ample
skin trowsers; and, as is the custom of the boors,
his pipe was seldom out of his mouth.   He is a
very civil man, and bears a good character for
kindness to his people.   His wife sat at the door,
in a close cap and blue cotton gown, sewing.
There were two or three long guns slung at one
side, and a pair of low stools were in the hut;
but neither table, chairs, nor bed; karosses, or
mantles of sheep skin, spread on the floor at
night, and rolled up in a corner during the day,
served the place of the latter; and when the
farmer gave the order to " Schenk een zoopjé,"
(pour out a dram), and then to " Skep op," (set
the victuals on the table,) a waggon chest was
drawn from one side, on which a cloth was spread,

and pewter plates arranged. Two large messes of boiled mutton were then produced, and boiled wheat, when, on the words "Kom zit bij," we placed our stools alongside of the chest, and each drawing a pocket knife, we made a vigorous assault on the viands, washing them down with warm milk, handed to us by a Hottentot female.

After the repast, we carried our stools outside the door, to a blazing fire in front of the hut, and sat conversing about country matters, till it was time to retire for the night.

Leopards and Boschmans are sometimes troublesome in this district. One of the former lately killed eleven horses here, before it was destroyed itself. Boschmans hovering about the frontier too, carry off a single sheep or a cow now and then from the flocks and herds in the field, which they kill and devour in some neighbouring dell among the mountains.

The Boschmans here, as elsewhere, have neither sheep nor goats, nor do they cultivate grain or melons. At one season of the year they catch with their dogs the fawns of the springbok; at another, the nests of the white ants are robbed of grass seed, and of the ants themselves, for food. Flights of locusts they delight in, and honey is sometimes most abundant; roots are found after rain by their green shoots; and in the months of July and August, ostrich eggs supply the wants of these " children of the desert."

When they visited Nieuwoud in their seasons of scarcity, he killed a sheep for them, and gave them a small present of tobacco, to prevent their robbing him.

I shot a fine blue falcon, which preys on snakes, in the Zwart Doorn river; and we also got several long-tailed finches and handsome doves for the collection. Suddenly I heard in the river the very loud report of a gun, and the whistle as if of a ball. I ran towards the sound, and found Elliot with the remains of his fusee in his hand, which was cut, as was also his lip. He had imprudently overcharged his piece, which had burst, and nearly destroyed him.

I now sent on two Namaqua messengers with presents of shawls to the chiefs at the Warm Bath and at Pella, on the Orange river, to announce my approach, to say that I came with no hostile intent, but merely to see the country, and to endeavour to open a trade with the natives, and I requested the chiefs to meet me at Lily Fountain, on the Kamiesberg Mountain.

We proceeded on the 7th of October to the Groene or Green river, where the heat was very oppressive—thermometer 95° in the shade of the waggon. At night, my people being comfortably housed in the tent, and myself in the waggon, sleeping over seven rockets, one hundred pounds of powder, and six hundred ball cartridges, lightning played about us, and thunder rolled over head, whilst rain fell in heavy drops.

We next passed by the Quick river (under the great Kamiesberg range), in which I had a most luxurious bathe; and before commencing the ascent of one of the offsets of the Kamies or Lion Mountain, the farmer Rouseau and his wife (encamped on the river's bank in the usual mat hut) liberally presented me with some loaves of bread.

Few have been so much indebted for hospitality and kindness, as I have been during my various wanderings. Travellers will generally find, if they attend to the usual forms of civility without cringing, and are affable without being too familiar, that their wants will be relieved, and attention will be extended to them; particularly by women, whose hearts are readily touched with compassion for those who are far separated from friends and home.

> " Ask the grey pilgrim, by the surges cast
>   On hostile shores, and numbed beneath the blast;
>   Ask who received him? who the hearth began
>   To kindle? who with spilling goblet ran?—
>   Oh! he will dart one spark of youthful flame,
>   And clasp his withered hands, and woman name!"

Snakes now become rather rife. Magasee one day was horrified by the appearance of one with, he said, a head as big as a tea-pot, rearing itself out of a bush as he passed, and glaring at him with its fiery eyes, and hissing with its fanged mouth. It was difficult also to avoid treading on the poisonous cerastus, or horned snake, which, a foot and a half long, and of a light brown colour, with dark spots, lay coiled among the stones in our path, and whose colour it much resembled.*

We reached a considerable elevation, and out-

* See end of chapter.

spanned at a beautiful spot called Hooge (high) Fontein, where we passed the night. On the morrow there was a most magnificent sun-rise among these grand mountain scenes. The blue and distant peaks rose like islands out of a sea of mist, which filled the valleys; the white veil was lifted upward with the increasing heat; and flowering bushes, and rocks covered with red lichens, were revealed in the foreground; and then the green and broad sides of the granitic mountains were laid bare. Thus the Great Luminary converted the mountains into mighty altars, from which immense clouds of vapour rolled towards Heaven, as if Nature were silently but most impressively offering a sacrifice of praise and adoration to its Divine Author.

We descended to a valley called the Two Rivers, where we found a community of Little Namaquas, belonging to the Wesleyan mission station of Lily Fountain. They were living in mat huts in a glen. The men were decently dressed in leopard skin or tanned jackets and trousers; and the women in sheepskin karosses and tanned petticoats.

We got milk for tobacco from the yellow-faced
and Chinese-looking Namaquas; and passing
through the valley, we again ascended by a steep
road to the higher parts of the Kamiesberg,
where we saw the strange koker boom, or quiver
tree, with its thick and silver-green trunk, hollow arms (from which quivers are made), and
leaves like those of the aloe. We passed through
level tracts, which were covered with crops of
corn; and on the 10th of October, a month after
I had left Cape Town, we reached the elevated
and beautiful Wesleyan mission station of Lily
Fountain.

CHAPTER III.

The Mission Station of Lily Fountain described — View from the Kamiesberg — Life on the Mountain — An Unpleasant Summons — Ascend the highest peak of the Mountain — The Cupper — Wax-seated Visitors — Abuses — The Post — Collection of Taxes — Grasping Functionaries — Anti-Mac-a-*dam* notions — Travelling Maxims — Charity and "never mind the clamour" — Kindness and Cruelty of the Boors towards the Coloured Classes — A Miscreant described — His rebellious language and injustice — He flogs and brands his Herd — State of Religion among the Boors — Anecdote of "The Sheep and the Goats" — Easy bearing of "an injured Husband" — Trials of Strength — Cobus Bulle, the Brandy Boor — A Wife shot by her Husband — Halcyon Days.

THE station of Lily Fountain is placed about 4,000 feet above the sea, and immediately under one of the peaks of the Kamiesberg. The highest peak, six or seven miles distant, and south, from the station, is estimated at 5,000 feet in height.* In the sloping mountain valley

* My mountain barometer was broken soon after I left the Cape, as Sir John Herschel predicted it would be; for he had never found those of the same construction to stand the least jolting or rough work.

D 3

of the station is found a good church, school, mission-house, and out-buildings; a productive garden, watered by an abundant fountain, which poplars overshadow; whilst around are the mat huts of the Namaquas of the station.

My worthy friend, the Rev. Barnabas Shaw, first formed this interesting station in 1817. He was succeeded by Mr. Edwards, who laboured here for fourteen years most successfully. Part of the remains of the Little Namaqua nation was here collected,— there are eight hundred on the books of the institution; and I was quite surprised and pleased to see the quantity of land they cultivate, stimulated as they are to exertion by the missionary, under whom are two corporals and six councillors, or heads of families, elected by the people by ballot.

The Namaquas of Lily Fountain had sown latterly about 100 muids, or 20,000 lbs. of wheat annually, and had raised from this 1,500 or 2,000 muids. Mr. Edwards was absent at Cape Town when I arrived at the station, and a thin-looking corporal (Buchas) received me. I thought that he was very poor from his appearance, and I in-

tended offering him the head and liver of a sheep
I was about to kill, to keep him from starving,
when I found, to my surprise, that he grew forty
muids of corn annually, had a span of fourteen
oxen, a waggon, twelve horses, and seven hundred sheep and goats!

Once a month disputes are settled here in
council, which are principally for cattle tres-
passing on corn land; and those brought before
the council can of course appeal from its decision to the field-cornet of the ward, or magistrate
of the district. Yearly a herd is appointed, and
yearly the ponds must be cleared out for the
cattle. For misdemeanors there is no flogging,
but a fine of goats is imposed. If honey-beer
is made, the maker of it is expelled the station,
and no native dances are allowed, for they open
the door to vice, the dancers being in the habit
of remaining to sleep where they danced, and
relations hearing of this, quarrels ensue. Thus
the missionary, besides having his spiritual duties
to attend to, the farming, carpentry, and smith's
work, has much to do with the temporal matters
of his people.

The winters are very severe on the Kamiesberg; snow lies there thick for two or three
months; and the people almost all go down the
mountain to the valley of the Two Rivers,
(where the missionary ought then also to live;)
but for eight or nine months the temperature
is delightful on the mountain: I found it 65°
and 70° generally at mid-day—with clear skies,
the perfume of wild flowers, the constant rustling
of the leaves of trees, and the notes of birds,
to soothe one in this retreat from the vanities of
the world.

Walking up the green slope behind the station
to a gorge between two summits of the mountain, I saw wave after wave of hills declining
towards the sea, over which a mist hung, and on
which, at night, an occasional ship conld be distinguished by its lights.   There were numerous
traces of rock rabbits about, and the Cape lark
whirred aloft, and dropped to the ground with
its melancholy note.

October 14.—Sunset by my watch at 6. 18.
50, observed variation 301°.

I intended to tarry on the mountain for a

short time to give the chiefs, for whom I had
sent, an opportunity of joining me here; and
I also meditated an excursion (by request of the
Governor) to inspect the mouth of the Orange
river, of which no account could be found at
the Cape. I therefore now practiced my people
with firing at a mark; and I also collected birds
and plants; arranged our stores for our future
progress; and I generally walked ten miles
daily by myself, scrambling among the rocks, or
through the deep valleys, in which were partridges and black and white bustards.

Returned in the evening, I sat down
solitary in a small room. A light made from
the fat of a sheep killed in the morning is
produced; the snuffers are a bullet-mould; a
pewter plate is placed on the board, and a flap
of wheaten bread with a calabash of rock honey,
an iron pot is brought in from the fire, into
which I plunge my fork, and produce either
a rabbit or a plump partridge. Moses, the
handsomest of my dogs (a black and white
spaniel), and the most sensible, whines at the
door; he is admitted to pick the bones. A

draught of churned milk finishes the repast: with my flageolet and books I conclude the evening.

This sort of life I tried to enjoy to the full, as I knew it could not last long, and I had now a strange summons to give up the expedition altogether.

It was now the 15th of October, and I received, by express from the Governor, an intimation, that having been now some time absent from my regiment, (though hitherto constantly employed on military duties), the authorities at the Horse Guards proposed to place me on half-pay, if I failed to join my regiment on the 1st of January, 1837. I was much disquieted by this announcement, occasioned probably by misapprehension of the manner in which I had latterly been employed, and to silence the complaints of those who did my duty at home. I saw my hopes of promotion in the service about to be annihilated; and now, having gone on so far with the expedition, and having put my hand to the plough, " I could not look back." I therefore wrote to say that I should risk losing my full pay sooner than now return to the Cape and to

England, which last it was hardly possible that I could reach before the beginning of the year; but, at the same time, I made a strong appeal to be saved from the threatened reduction, and I urged my private friends to use their influence to save me; and I found, a year afterwards, that it was then fortunately spared, though I prosecuted the journey in the supposition that I had been put on half pay. To try and dissipate the effects of the unpleasant announcement from the Cape, I rode off to Ezel Fountain, field-cornet Engelbreght's place, five miles south of Lily Fountain. I found there the field-cornet's son-in-law, M. Rouseau, whom I had before seen, and with him I ascended the highest summit of the Kamiesberg.

Everywhere around us were masses of mountain—we neither saw houses, trees, nor water anywhere, for a mist lay on the South Atlantic, and there was no "shining river" flowing through the wilderness: rocks of red and white quartzose rock were under foot, among which grew red everlasting flowers. South of us was a conical mountain perfectly bare, with a crown of loose rocks, and on its sides the strata appeared most distinctly inclined at an angle of 45°, and dipping to the east, where the karroo, or dry plain, was faintly seen.

I returned to Lily Fountain, where I found Taylor laid up with severe pains in the face, called zinkins in the Cape, arising from cold. I tried all sorts of remedies without effect—hot water, laudanum, &c., and he got no relief, till I made an old Namaqua cup him with a small horn, which he applied to his scarified cheek, and sucked at the small end till the blood flowed. The Namaquas are very fond of local bleeding, and the backs and sides of many of them are thickly scored with the knife. They are not over particular about having a horn, the mouth itself being frequently applied to the wound.

At Lily Fountain one great source of annoyance to me was the lengthened visits of the boors. I was very glad to see and to make acquaintance with the farmers, from whom I received much hospitality; but it was too much to expect that my whole forenoon should be consumed in answering, at long intervals, such questions as this:

If I was not afraid to travel alone;—if the Governor was "versch," (pronounced fresh), or in good health;—if the king was an old carle; —how many children the Governor had;—if they were married or single;—if I was tired of this place?

These, with their answers, would perhaps occupy an hour and a half; the lusty boor all the time sitting in the middle of the bed, for want of a chair; while I, in my turn, would ask if the farmer was married;—the number of *his* children;—would repeat the old story of the rust in the corn;—the horse sickness, and wollig schaap (Merino sheep);—asked how far it was from one place to another;—where the game was to be got: —then would hint how much I had to do;— shewed my writing materials, which would merely produce a drowsy "yaw;"—would look at my watch, and, on being asked the time, hear, as usual, the reply of, "het iz niet laat" (it is early yet);—and, as a *dernier resort* to get my wax-seated visitor out of the room, I would be compelled to put on my hat, with an apology for being under the necessity of visiting a patient.

" Zo," says the boor, " I'll go with you, Mynheer;" and once up and out, he goes to look
after his riding-horses turned out to graze, one
of which is perhaps (fortunately for me) lost;
which affords me some hours' quiet, until I give
him his " bread and cheese," and see him fairly
in the saddle, with "groetens ta huis" (compli-
ments to all at home).

Luckily sometimes Cape papers would arrive,
in Dutch and English, and after the usual en-
quiries were duly answered, I would place the
papers in the hands of my visitors, to spell
at their leisure, and thus be able to go on
with my writing without a breach of good
manners.

I found here, as elsewhere at a distance from
the seat of government, that there were many
abuses to correct. Thus, a letter commonly took.
four months to travel between Cape Town and
Lily Fountain, having taken a long rest at each
of the field-cornet's houses before it was sent on.
A letter which had been gone for three months
from the neighbourhood of Lily Fountain, was
returned to the writer, an Englishman, because

he had written on it Cape Town, instead of Kaapstad!

The field-cornets receive two hundred rix-dollars (15*l.*) a year, and are exempted from taxes, though "all the world" else is taxed; and yet the manner in which the opgaaf in this district is collected is most objectionable and oppressive. Instead of the field-cornets collecting the taxes, and sending, or taking, the amount to Clanwilliam, all the heads of families, white men, Bastaards, or free Hottentots, are obliged to go to Clanwilliam personally, in the month of April, the driest time of the year, (when the grass is burnt up and the water scarce), with the amount of their opgaap, whether it be thirty dollars or five. Thus valuable time is lost on the road, the cattle suffer severely on the journey, though some people are obliged to walk, for want of horses, and all leave their families without proper protection from vagabonds.

The field-cornets are not in the habit of sending round the Government Gazette to the farmers as they ought to do; thus the people are kept in ignorance of the ordinances, and are not aware of the new laws which may be framed for them to obey.

But the greatest injustice is yet to be told. One field-cornet had actually appropriated to himself no less than eight places, and to five out of the eight he had no right or claim. It would have taken a man two days, with two good horses, to have ridden round this functionary's land. The following case of a certain field-cornet and his sons, exhibits the manner in which these gentry sometimes procure their places. An old farmer had occupied a loan place from government for twelve years. He died in the beginning of 1836, leaving a son Erasmus, who was lame, and who had no other means of subsistance than grazing cattle and sheep on the farm. Girt, the field-cornet's son, comes and turns *his* cattle on the place; Erasmus complains to Girt's father, who says that Erasmus must leave the farm: that it was only a loan place; and that he, the field-cornet, being a government dienaar (officer), can do what he likes with the land.

But the usual mode resorted to here, and in other parts of the colony, to get land is this.

A farmer pretends that there is no water on his place, but that over the hill there is a fountain. He accordingly gets or takes the loan of the land about this, and thus excludes other settlers who might wish to locate themselves in the ward. I was provoked with one farmer, an occupier of many places, saying, even at the foot of the Kamiesberg, where there is generally plenty of rain, and consequently no want of water for cultivation or grazing,—"Mynheer, what do you think of the country? You see we have no water here."

"No water!" I replied, "there is plenty if you would only take care of it. But I see no *dams* here."

"Dat is waar (that is true) Mynheer; but Cobus (somebody) made a dam once, and the water ran all out through the sand below."

"Are there not ant heaps here of fine clay," I answered, "to plaster the bottom and sides of the dam, to make it hold water? Is there not an Englishman of the name of Kennedy who set to work in this neighbourhood with a single Hottentot, and made a dam in two days with stones, clay and bushes, at the place of his

father-in-law; and which dam holds water all
the year round?" (This occurred two years
ago, and while the work was going on one or
two of Kennedy's Dutch connexions stood by
with their hands in their pockets, and never
offered to take off their jackets and assist him.)

" Dat is ook waar Mynheer (that is also true,
sir), but all this is zeer moelijk (very troublesome.)"

I dislike the traveller who indulges in indiscriminate censure of any class of people; but I
have also no great respect for the voyageur who
pusillanimously conceals all the faults and failings
of the people among whom he travels, fearing
that he may afterwards be called to account if
he reveals them. Charity is a divine injunction,
and the hiding of our neighbours' faults; but
surely by this it is not intended that we should
withhold from noticing them in all circumstances. Let us trace them to their sources,
making every allowance for the situation of the
parties with whom we are dealing; and let us,
if we can, suggest remedies.

The taint of slavery here, as elsewhere, makes

the white man lazy. It is considered a disgrace also, to be seen to do what is usually the work of the coloured races. Had we been born and brought up among the boors we would doubtless have thought and acted exactly as they do; but, thank Heaven! slavery is now at an end in the British dominions; though, as far as regards the Cape, emancipation may not have been so judiciously carried into effect as it might have been, and with justice to the slave owners (but of this more hereafter); and it is anticipated, that if a thorough and sifting investigation into the rights of property is made in this and other districts of the Cape Colony, and those who have no right to farms are deprived of them, and the land given to other Dutch, German, or English settlers (it matters not who they are, if they will only turn what they get to account), that the colony generally will assume a new and an improved aspect.

Though some of the farmers are kind and indulgent to their people, others treat them with the greatest severity. An example of the first class is Arnoldus Vanzyl of Keerom, who is, perhaps, too indulgent to his people, and they therefore sometimes take advantage of his good nature, whilst every wandering Hottentot is taken in by him, and his wants relieved. I forbear to mention the names of those who treat their people with injustice and cruelty, that the feelings of the descendants of these men, who may not tread in the steps of their parents, may be spared; but I state facts, though I suppress names.

. One farmer will not allow his Hottentot shepherds to sleep during the winter's frost and snow in mat houses: they lie out unsheltered, and only defended from the bitter blast by a few bushes placed to windward of their lair. " For," says the compassionate Boor, " if the schelms (rogues) were to sleep in huts, they would let the sheep all run away."

One of the worst characters, and perhaps the most mutinous and disaffected to the English government, in the sub-district of Clanwilliam, is a field-corporal. He is loud in his abuse of the English, even before Bastaards and Hottentots. He and some of his companions, when they get together, are principally occupied in complaints such as these. " Ver doem de government! it presses us in every way; we de armen boeren (the poor farmers) pay for every thing. And now we cannot lift a hand to a Hottentot baviaan (ape) without having to go before the magistrate for it. In the old times we could do what we liked with them, and no one meddled with us; now, with the government and the zendelings (missionaries), we can get nothing done—Der duivel !"

Yet, notwithstanding this abuse, and "the poor Boors paying for every thing," the taxes are exceedingly light—say, thirty rix dollars or 2l. 5s. (the price of an ox) for a place of three thousand morgen, or six thousand acres, which supports many hundred head of cattle, horses, and sheep, is surely moderate enough.

An instance of the disgraceful conduct of this field-corporal may here be given. A few years ago part of a broken tribe of Caffers wandered from the south-eastern parts of South Africa as far west as the northern boundary of Clanwilliam; and there sitting down on an unoccupied piece of land, proceeded to turn it up with their rude wooden spades, and to plant a little corn with a stick. Though these people were quite harmless and peaceable, and had settled at a distance from any farmer, yet the field-corporal mercilessly ordered them off the land; and because they did not move when he desired them, he made his herd drive his goats on the land when the corn was ripe, and eat it off. An old Caffer, watching the corn, remonstrated; but it was of no avail. The field-corporal actually sowed corn next year on the land which had been cleared by the poor Caffers, who retiring to another spot, and the persecution of this ruffian still continuing, he wrote to Clanwilliam (after the war with the Caffers on the eastern frontier had commenced), to the effect, that a body of Caffers had come into the district, and that he proposed to drive them out by force of arms, to prevent their injuring the farmers. On Mr. Ryneveld reporting this to government, he was sent up to the spot to inquire into the true state of the case; and on ascertaining that it was only a diabolical scheme of the field-corporal to get rid of the unoffending natives, he was ordered to allow them to remain on their location. ·

The shepherd of the field-corporal, a Bush boy, in returning home with the flock one evening through a kloof or pass, stayed behind to bring up some of the lame, when a panther springing from behind a rock, destroyed two or three of the foremost sheep. The field-corporal on counting his flock missed some, and on being told by the boy where they had fed that day, he went on the " spoor" (or track), and finding the carcases, he proceeded to flog the boy most unmercifully, and then scored his back with a hot iron. The boy fled to his kindred in the wilderness, and showed his wounds, the effect of which on his people may be well conceived.

I said before that the farmers of Clanwilliam labour under the great disadvantage of not having had for years the benefit of the progress of a clergyman among them. It is therefore not to be wondered at if many of them are careless of religious observances; indeed it is a matter of surprise that any of them keep up the outward forms of religion at all, living as they do so far

E 2

apart from each other and from the world. Some do attend however to family worship, and pray to their Maker, and praise him in the midst of their wives and children; though hardly any of them admit their coloured servants to the house on these occasions.

Thus a missionary told me that he had halted at a farm-house on a Saturday, intending to remain there over Sunday, so as not to be on the road on the day of rest. On the sabbath morning he proposed to the farmer to assemble the people and hold service, which he assented to, and called his wife and children.

" Where are the Hottentots?" asked the zendeling.

" The Hottentots!" cried the farmer, " you would not have them with us also? We are told in the Bible that the sheep are to be separated from the goats, and I cannot therefore admit the Hottentots."

" Very well," said the missionary, " as I am sent to teach all mankind the way of salvation, I cannot consent to hold worship unless white and black both join in it."

The farmer would not agree to this arrangement, on which the missionary very properly went out to his waggon, and calling his driver and leader, he prayed with them. Presently he heard the people in the house singing a hymn, and then the door opened and the farmer halloed to his people to bring the horses, and tread out the corn on the threshing floor. And thus was the sabbath spent!

Though the farmers affect to have a great abhorrence for any admixture of black blood, yet, strange to say, I saw, at a farm house, several dark children running about, who, I was told, were the offspring of one of the daughters of the family by a Hottentot youth. Another of the daughters of the same family married a Boor, and, seven months after marriage, produced a black child, which a trader seeing, asked " Hoe kom dat ?" (How did that happen ?) When the husband coolly replied, " that one day his wife was going out and was frightened by a black man, whom she suddenly saw behind the door, and that the child became black in consequence." The wife was by, and on hearing this she merely laughed.   So both parties " thought no harm."

On the morning of the 18th October the thermometer, at sunrise, at Lily Fountain, was at 40°.   The day was rainy and cold.

Besides practising my people with shooting at a mark (with some loose powder and balls I could well spare), I also made them amuse themselves with running, leaping, wrestling, and other feats of strength, to keep them employed, and to exercise their muscles.   The Namaquas have two or three odd ways of trying one another's strength besides wrestling, of which they are very fond: thus, one lies flat and stiff on the ground, whilst another tries to take him up between his legs, clasping his hands behind the back of the one lying down, who, all the while, makes himself as heavy as he can.   Another trial of strength is thus: one man ties a rope diagonally across his body (over one shoulder and under the opposite arm), he then goes down on his hands and knees on the grass, whilst another, holding the end of the rope, tries to prevent the first creeping along the ground.   A third way, was for a stout fellow to lie with his face on the ground, whilst two others would place their legs across his back, and hold by each other's legs; the first holding the two with his arms, would then endeavour to rise and walk off with them. Henrick, my driver, proved, at these feats, to be the champion amongst the coloured people who competed with him. I found him also to be a most invaluable servant—very attentive, quiet, good natured, and respectful.

A "brandy-boor" now came on the mountain; that is, a farmer with a waggon-load of wine and spirits, who tried to dispose of "liquid damnation" (as a fanatical person at the Cape used to term brandy) to the Namaquas. This Boor was commonly known by the *soubriquet* of Cobus Bulle, as he was a big, red-faced man, with very coarse manners. He sold his wine for a rix-dollar, or eighteen-pence the bottle, and his brandy for two shillings. I gave a hint to the corporal to look after the people, and prevent their drinking at the brandy waggon; and, accordingly, Cobus Bulle wended his way from the station, complaining of the badness of the roads, of the cold of the mountain, and, above all, of the

armoedigheid (poverty) of the people, which prevented their dealing with him.

The brandy-boor was succeeded by a sturdy
smith and field-corporal, 'ycleped Vander Westhuys, a good and civil man, who brought with
him a young Bastaard from the boundary, who
had just shot his wife. As this occurrence had
taken place whilst the Bastaard was temporarily
grazing his cattle beyond the frontier, Vander
Westhuys was uncertain what to do with the
prisoner. The case was shortly this:—the Bastaard had been married only ten months, and
had lived happily with his wife; his waggon was
packed and ready to return again to the colony,
when, on asking for his sambok, or whip of
rhinoceros' hide, his wife pointed to the ground
where it lay; the Bastaard took it up, and pretended to chastise her with it; she ran at him,
in a feigned passion, with a ladle, to attack him;
he snatched his gun from the waggon, pointed it
at her, when it went off and lodged the contents
in her side! The poor girl expired in an hour.
I recommended the field-corporal, instead of
sending the Bastaard a long journey to Clan-

william, to send him to his father, a boor, and make him answerable for his appearance, and in the meantime to report the circumstances of the case to Mr. Ryneveld, at Clanwilliam.

Thus I spent my time at Lily Fountain, variously employed during the day, and occupying my solitary chamber at night; and I enjoyed great calm and peace of mind during " these halcyon days, far too serene to last."

" Non numero horas, nisi serenas."

CHAPTER IV.

I was preparing to set out alone for the
mouth of the Orange river, when a favorable
opportunity presented itself for my visiting it.
Two of the Rhenish missionaries had arrived in
the neighbourhood, and they wrote me to say
that they proposed going to the mouth of the
Great River, and to Pella, to look out a site for a
new station, and that having got a horse-waggon,
I might find it convenient to accompany them.

ARIS, ORANGE RIVER.

Etched by William Heath.

Published by Henry Colburn, Great Marlborough Street, 1838.

Accordingly, allotting certain occupations to my people in furtherance of our journey, (such as collecting birds, procuring meal and sheep for the road, arranging the baggage, &c.), I left my men comfortably housed on the mountain, and riding with a guide by J. Coetzer's farm, from the proprietor of which I experienced much kindness and liberality, I descended the steep, rocky, and long pass of Cardow, and arrived in the low country, at the widow Vanderkniver's, above whose residence a high peak is twisted like a corkscrew. Next passing through valleys and between hills, at mid-day I "off-saddled" for a couple of hours to refresh at the mat hut of field-cornet Agganbag. I found his three fresh and strapping daughters boiling soap, prepared with fat and the branches of the soap-bush.

A large black snake which had crossed our path, led us to talk of Namaqua poison doctors, who pretend that they possess a charm against the bad effects of snake bites, by catching poisonous snakes, squeezing the poison out of the bag at the bottom of the fangs, swallowing

this, and then throwing away the snake.   There
are some people also who allege that they inocu-
late themselves and their children, to render snake
bites harmless, by cutting the skin and applying
snake poison to the wound; while the grand
remedy of these poison doctors, when called to
assist a person who has been bitten by a snake,
is to wash their own greasy cap, and to give the
patient the water to drink: this is doubtless a
sickening dose !

A small air-piston with a glass cup; sucking
the wound from a snake bite, opening it with a
pen-knife and exploding a little gunpowder in
it; *eau-de-luce* applied externally, and taken in-
ternally in a little water, to prevent coagulation
of the blood; drinking a great quantity of milk;
or olive oil applied hot to the wound, and taken
inwardly ; are all good remedies against the bite
of poisonous reptiles.

Journeying westwards from Mr. Agganbag's
field-hut, I was met on a hill side by Arnoldus
Vanzyl, Mr. Archer, a trader, and Mr. Anderson, lately the master of a merchant vessel.
We rode on to Keerom, or " turn round, "

as a hill prevents waggons going further to the north without a great detour.

I tarried at Keerom for the night, and was hospitably treated.

I now ascertained the existence of two new objects of interest on the map. The first is a small but productive seal island, between the Orange river and the Kowsie or Buffalo river, (the last the boundary of the colony). To this rocky island the Namaquas swim from the mainland, from which it is not far distant; and, in the months of November and December, they find abundance of seals there, for the purpose of breeding. The old ones will not leave the island as long as the whelps are on it, and are thus knocked on the head with six-feet poles. In the end of 1835 two traders (Eddington and Kennedy), with the assistance of the Namaquas, had got between four and five hundred sealskins off the island. These the Namaquas willingly gave up for five or six shillings each. They sold in the Cape for eighteen shillings; and in England from two to three pounds is got for good seal-skins. The natives dry the flesh of the seal, and subsist on it.

Four or five of the Namaquas were lately drowned by being carried out to sea with the current whilst attempting to swim to the island, after this they were glad to have the assistance of Europeans to make a raft, and assist at the capture of the seals.

The next discovery of importance in this neighbourhood, which I shall now notice, is that of a new bay called Rooé (red) Wall Bay, about forty miles to the south of the Zwartlintjes (black ribbon) river, and close to the mouth of the Spook (or ghost) river.

The north-western part of the colony labours under the serious disadvantage of being at a great distance from a market. Three weeks or a month are usually consumed on the road between Lily Fountain and the Cape with an ox-waggon; and the roads in many places are sandy and bad. The farmers therefore in the Bok-keveld, and other wards of Clanwilliam, seldom attempt a journey to the Cape: and though they may raise corn, they cannot dispose of it; they therefore content themselves with getting clothes for themselves by exchanging their cattle and horses with one or two traders, (who frequent this part of the colony), for goods; and by growing sufficient corn for their own consumption. Thus they do not go much beyond " clothing for their backs and bread for their children."

It will therefore be immediately seen how the condition of the farmers would be improved by their being able to export produce by sea from their immediate neighbourhood, by which means, also, greater supplies would be thrown into Cape Town; and the farmers would obtain the necessaries they require at a much cheaper rate than they do at present.

For some time there had been a rumour that there was a bay on the west coast of the colony, near to the frontier, which might be turned to account; but no one from the Cape had proceeded to examine into the truth or falsehood of this; when I found that Mr. Anderson, before mentioned, had come up privately, in the employ of two or three Cape traders, to search for this bay, and he had discovered it, though, until I made it known at the Cape, it was kept secret.

Rooé Wall Bay derives its name from lofty cliffs of red sandstone which face the sea at its eastern extremity. It is an indentation in the coast line of less than a mile in length and breadth; the entrance is broad, and across it blows the two dangerous winds at the Cape, the N. W. and S. E., so that it is sheltered, except from the western swell of the South Atlantic. Precipitous cliffs are on the two sides of the bay, and two sandy beaches at the bottom of it, separated by a mass of rocks. These sandy beaches are favourable for hauling the seine, and abundance of excellent fish are to be procured here; such as the delicious Roman fish, Hottentot, "Jacob Fever," mullet, stump nose, and clip fish.

There is one danger on entering the bay; viz., a rock on which are only two fathoms of water, and twelve fathoms all round it:—a buoy will easily point out this danger. At the entrance of the bay the depth is twelve fathoms, which decreases to six, four, and three and a half. In five and a half fathoms there is good anchorage in the middle of the bay, with sand and shells. All about the bay, the land is covered with shrubby plants, and there is grazing for cattle:

the soil is sandy, and drinking-water is to be procured by digging on the beach. There are
plenty of shells for lime on the coast, and abundance of drift wood about the mouth of the Orange
river.

Rooé Wall, therefore, affords every possible
facility for shipping produce, and also for salting
provisions and establishing a fishery; and it is
earnestly to be hoped that it may be the means
of " opening up" the " section" of country in
which it is situated.

Leaving old 'Nolus Vanzyl enjoying a *kettle* of
" tea water" (for he makes a boast of being able
to drink this beverage " geheel dag," from
morning to night, " at no allowance,") and
sweetening it with a foot square of honeycomb,
I rode by Wild Paard Hoek (Zebra's Corner,)
and the corn fields of Komakas (red clay,) and arriving at the London Mission station of that
name, I was kindly received by the worthy
old German missionary the Rev. Mr. Schmelen.

Komakas consists of a long mission house of
one story, a church, and outbuildings, situated
under a mountain of about a thousand feet high,

and facing the south, and some rocky hills. In the valley, and opposite the buildings, were the mat huts of the Bastaards and Namaquas of the institution. To save the grass about the station for another season, most of the people of Mr. Schmelen were in the field with their flocks and herds, and only about thirty or forty were now present.

There was a small wind-mill for grinding corn, also a good garden; and no less than five fountains of excellent water were in this green and secluded valley, in which the distant roar of the sea can be heard, and over which peace seemed to wave her olive branch.

> Sweet peace of heart, from false desire refined,
> That pours Elysian sunshine on the mind,
> That soars with strong and steady flight sublime,
> Where disappointment never dared to climb.

Mr. Schmelen had laboured for upwards of thirty years in the wilderness of Great and Little Namaqua land, and in the region of the Orange river, principally. No one can be more highly respected than he is by the natives; among whom he has been very successful as a teacher, and over whom he has great influence. Single and solitary, he wandered about with the people, living often on game, and without bread, for a great length of time. He established the station of Bethany, occupied it for some years, and at last was forced to abandon it, as shall afterwards be recounted: he had travelled further to the north in Great Namaqua land than any white man previously.

Mr. Schmelen is compactly made, and combines great energy with excellent judgment and good nature. His first wife was a Great Namaqua woman, who led a most exemplary life, and by whom he had several children; his second wife is from the Cape, and is most active and indefatigable as a school-mistress. May they both long labour in their sphere of usefulness.

I found the Rhenish missionaries, Messrs. Lepold and Terlinden, at Komakas. Our few arrangements were soon completed for a visit to the mouth of the Orange river—such as laying in a small supply of bread and dried meat, in skin bags, rolling up our sheepskin karosses or mantles, to sleep in, and placing them, with two or three fowling pieces, in a light horse waggon.

We left Komakas on the 27th of October, with eight horses, and passing round the western extremity of the mountain, we journeyed north over a flat covered with bushes, and with mountains on our right hand. We went on at a rapid rate, old 'Nolus Vanzyl holding the reins, and an active Hottentot the long bamboo whip. Occasionally the Hottentot jumped off to flog the sluggards, and ran alongside of the waggon for some distance when the horses were at a gallop, and then sprung on the waggon again. We did not travel on a railway, and in a horse waggon it is rough work for the bones when the team "goes ahead."

After five hours ride, we arrived late at the Kowsie river, the boundary of the colony, and passing through its dry bed, and the mimosa and dubbee, or tamarisk trees which lined its banks, we outspanned for the night on the north side of the river, at a place called Bont Köe, or Brindled Cow. We turned the horses out to graze in the dark, collected some dry bushes to make our fire and boil a kettle, and after a light supper, eaten whilst seated on logs and small water kegs, a hymn was sung, and kneeling in the sand, thanksgiving and prayers were devoutly offered up. After which, shaking the sand out of our shoes, and putting them on again, for fear of hungry dogs or hyenas carrying them off in the night, we cleared the stones from a lair of sand, and rolling ourselves up in our sheepskin karosses, without, of course, undressing, we slept uninterruptedly till daylight.

We found the horses by their "spoor" some distance off, and inspanning, we pursued our evening journey, and saw the traces of ostriches and zebras. After four hours ride we outspanned. On the following day we lost the horses for many hours, as they could find no water, and therefore had strayed in search of it.

Three hours ride took us to Ukribip (Scratch Claw Place), where beside a pool of water lay two or three families of Little Namaquas in mat huts.

Whilst lately staying for a night at a farmhouse within the colony, but near the borders,

I had been rather amused with a white man coming into the room where I was writing, and with some preparatory hems, and twisting of his hat, saying, " Sir, I am sorry to interrupt you."

" Never mind, can I assist you in any way?"

" Yes, sir, I think you can."

" Tell me in what way."

" Well, sir, I sent in a memorial some time ago to government for two places, where I think I could make a living: they are named Ukribip and Nubip: they are a little way beyond the Kowsie."

" Why, that is beyond the boundary !"

" There are other farmers on places beyond the Kowsie, and I hope that I also may get the places I have applied for, for I have a wife and two children, and Ukribip is only a few hours from the Seal Island."

" And what, pray, would you do with the Namaquas·who are now living at Ukribip ?"

" Why, they are lazy, and must just move further north, to the Orange river."

This is a sample of the little regard which is

paid to the rights of the aborigines by people who have been born and bred to believe them merely denizens of the soil at sufferance.

With fresh horses, procured at Ukribip, we galloped on at a rapid rate, and ascending an eminence, found ourselves at a Namaqua hamlet called Kama (water place), where were about a dozen huts among the hills.

The people here exhibited the old dress of Namaqua land. Many of the men wore a leathern girdle, from which in front was suspended part of a jackal's skin with the fur outward, whilst behind dangled a square piece of stiff leather. conical fur caps were on their heads, a karosse or mantle of sheepskin depended from their shoulders, whilst sandals or buskins of untanned leather were on their feet. In their hand they commonly bore a jackal's tail on a short stick, and with this Namaqua handkerchief they brushed the perspiration or dust from their eyes and face, and then dexterously twirled it between their palms.

The men were thin and athletic, of an olive-brown complexion, and with short noses, pouting

lips, and narrow but keen eyes; their general
height was five feet six or seven inches.

For arms they had some old muskets and long
guns obtained from the colony, (for four, six, or
eight oxen each piece) besides assegers or javelins,
composed of a slender shaft, five feet long with
a small blade of iron inserted into the upper
end, which was bound round with leather—a
knobbed stick to throw at game, with which they
are very dexterous—and bows and arrows; the
former is about three feet long, and is strung
with the back sinews of deer, the latter are
composed of a reed shaft, into which is inserted
a polished piece of sharpened bone, which is
usually surrounded at the point with a black waxy
looking substance; this is poison prepared with
gum from the milky sap of the euphorbia, and it
kills the game without destroying the whole-
someness of the flesh; occasionally a few of the
arrows have a barbed head of iron. Two dozen
arrows are contained in a case of leather, or of
the koker or quiver tree (aloe dichotoma.)

The women wore skin petticoats, or the Namaqua broek karosse, consisting of a prepared

sheep or goat skin, so arranged, as to depend from the waist in a broad oval flap behind, and and in front to be only a few inches in depth, where also a tortoise shell with a long fringe of leather thongs was suspended; this bunch of thongs reached to their ankles, and with it they sometimes chastised the children. The breasts were uncovered, strings of porcelain or glass beads were upon the neck, the woolly hair of the head was carefully concealed with a striped cotton handkerchief, though further in the interior a covering of softened leather is used; from the shoulders hung an ample sheep skin mantle ornamented at the nape of the neck with a square piece of leather, on which, black and white chequers of goat skin were sewed. They took off the kaross when employed in any hard work.

The young women were plump, good looking, and fairer than the men; the elderly females were invariably much wasted. The children were naked and looked healthy.

The general fare of old and young was the milk of their cattle or goats collected in cylindrical wooden vessels, called "bambus," they had often only this simple fare morning and evening for weeks together; they are reluctant to kill their sheep unless they meet with some injury, to kill a heifer is a rare occurrence, so that if not successful in procuring game, such as the eland, springbok, or ostrich, they content themselves with milk, a lump of gum from the *acacia capensis*, or bulbs and roots discovered by their leaves appearing on the surface after rain.

The huts were the same round-topped and circular-mat-covered ones which we had seen the boors use on their spring pastures within the colony; and their furniture consisted of little else than a few skins, whilst the bambus were suspended from a forked stick opposite the low door.

There was no sign of cultivation round the hamlet of Kama, though sometimes a little tobacco is grown by the people, and melons raised. The value of tobacco in Namaqua land is such, that for it many of the men would not hesitate to give up their wives and daughters;

and a roll of it, some distance from the colony, will purchase an ox.

The pipes of these Namaquas were composed of green serpentrine from the Kamiesberg, and were straight tubes three or four inches long, narrowing at the mouth piece, and not broader at the other end than to permit their insertion into the common brass tinder box. These pipes were neatly turned and ornamented with a little carving.

> " Little tube of mighty power,
> Charmer of an idle hour!
> Happy thrice aud thrice again,
> Happiest he of happy men,
> Who, when again the night returns,
> When again the faggot burns,
> Can afford his tube to feed
> With the fragrant Indian weed ! "

The Little Namaquas are a good people, they are neither vindictive nor blood thirsty; like other barbarous people they are quite regardless of the value of time—the men lie in the sun when not driven to the field to procure game— whilst the women make and mend clothes, and milk the cows and goats. The Little Namaquas are sensual, and have two or three wives if they can afford to keep them—though the missionary does all he can to prevent this sinful practice. Through him also, they have a knowledge of religion. Though dancing is discouraged by the missionary, yet both it and the drinking of honey beer is practised privately.

The Namaquas put up a scherm, or screen, for us of boughs and mats, and with a fire at our feet, we lay there comfortably. On Sunday the 3rd, the people were assembled round the scherm, divine service was performed in Dutch, and was translated by an interpreter into the clicking Namaqua language.

We ate here the thick and reddish root called *canna*, the more slender are called *quibé*, and some bulbs which, when roasted, formed a good substitute for bread.

For four months in the year, (from July to October, inclusive,) the people lie at Kama; the rest of the year is spent at the Great River. In all 1835, owing to the drought, there was not a soul between the Kowsie and Orange rivers. The people living between these rivers, amounting only to a few hundreds, would have no great objection to be placed under the colony. They could not, of course, pay taxes in money, but could give a fat ox annually, in token of allegiance to Her Majesty. It has always been the object of Mr. Schmelen (under whose charge these people consider themselves to be,) to impress on the minds of the Namaquas the necessity of conducting themselves as if they actually lived in subjection to the colonial laws; and they have, therefore, a very salutary respect for English authority. If placed under the colonial government, the people would be more under the control of the missionary; and of course no white man would be allowed, on any pretence whatever, to use their watering places or occupy their grazing grounds.

The presence of a resident magistrate, altogether unconnected with the farmer, is particularly required about the Kamies mountain, or at some other convenient place near the frontier, to protect the Namaquas who might be inclined to come there and barter their cattle for goods.

Many of the Namaqua tribes are very rich in
cattle, which they would willingly barter within
the border for cloth and cutlery; but they are
afraid at present to venture into the colony
without being adequately protected. I asked
one or two of those living about the Orange
river why they never took their cattle into the
colony, but preferred going to Angra Piquena
Bay with their herds; before reaching which
they often suffered most severely from thirst on
the road, and when they did get there, they
were often grossly imposed upon by the whalers;
obtaining only two quart bottles full of coarse
powder, or forty bullets, for an ox; and even
sometimes being made drunk, and getting nothing at all for their property. To this they
replied, that they had tried once to take cattle
into the colony, but that the first farmers they
met abused them—asked them whose cattle
these were they had stolen; if they had been
plundering the Damaras; and said, " *Vordoem de
Hottentots!* what business have they with cattle?" So becoming afraid of violence, and
seeing they had little chance of fair dealing

with the white men, they had never ventured to the borders of the colony again.

One cannot conceive a more dastardly and selfish spirit than that which could induce white men to behave in the manner that some of the whalers do. The natives wish to deal fairly, and part with their property, in their ignorance, for the value of a few pence; and, not content with getting them on these terms, the captains and crews of some whalers actually rob the natives; careless of the bad effect which this conduct will have in future dealings between the ships and the Namaquas.

On shore everything is promised: but when the natives are induced to go on board with their cattle, they are either frightened into parting with them for next to nothing, or they are made drunk, and sent on shore without any remuneration; and still, with all this, they prefer Angra Piquena to the colony.

I said, "a magistrate unconnected with the farmers is required on the borders." At present the magistrate at Clanwilliam is too far distant, and the field-cornets and the farmers are all related or connected: every one is oom or neef (uncle or nephew) to his neighbour, so that it is not very likely there can be much justice got out of a field-cornet, on the servant of his nephew complaining of ill treatment. Besides, most of the old farmers cannot get over their thorough contempt for the coloured races. It is, therefore, evident, that unless the magistrate is a man without local connexions and prejudices—unless he does what is right, without minding the breath of popular applause—the *vox popularis auræ*— he will do little good in this quarter. But that a good and upright magistrate is necessary here is very evident; particularly if Rooé Wall Bay is made available, and there is a brisk cattle trade in Namaqua land.

We left Kama, and in seven hours arrived at a place called Jackal Puts or Holes, where we slept under the lee of the bushes, and before dawn we were awoke by the uneasy yelping of a troop of "lion's providers," which came close to us, though we could not see them. Most of the dogs of the country can catch the jackal, one species of which is exceedingly handsome, with reddish sides, whilst down the back is a broad black and grey stripe: fourteen skins of this sort of jackal make a beautiful and valuable kaross.

Doorn poort (thorn pass)—two hours—was in a chain of mountains in continuation of the Kamiesberg range, and then three hours through heavy sand, brought us to Aris, a Namaqua village of about twenty huts, where we first saw the great Orange river.

CHAPTER V.

Sketch of the Gariep, or Great Orange River — Its Beauty
and Utility — Why White Men are likely soon to sojourn
on its Banks — Abundance of Copper, Iron, Ebony, Gum,
Bees'-wax, &c. are to be found there — Floods — Hippo-
potami—Awkwardly situated — Alligators — The Author
sits in Council at Aris — An attempt at gross imposition
on native credulity — Leave Aris for the Orange River
Mouth — Immense Flocks of Wild Fowl — Game — The
Mouth of the River — Can it be entered ? — A Boat Bay
—Interesting Discovery of Copper — An Orange River
Copper Company proposed — The String Gun — Leave
Aris — Ostrich Chase — Habits of the Ostrich — Its
Enemies — Ukribip — The Kowsie — Heat — The Se-
cretary Bird — Komakas — Keerom — The Gemsbok —
Lily Fountain — The Rev. Mr. Edwards and the Chief
Abram — Mission School — Story of an Old Woman—
Merinos — English Instruction for Native Children re-
commended — The Robber Stuurman — The " Shrik-
kelijk reis."

THE Gariep, or Orange River, is the most
important natural object on the map of South
Africa.   Traversing the great continent from
east to west, divided in its upper course into
three considerable branches, the Nu, Ky, and
Maap, (the black, yellow, and muddy) Gariep,
its waters roll from a lofty mountain range to the

ocean, over a distance of upwards of a thousand miles.

All those who have had the good fortune to see the Gariep agree in praising its beauty. Its broad stream at one time rushes· tumultuously over a rocky and shelving bed, then is spread out into a translucent lake, then is hurried over a rock four hundred feet high, forming a grand cataract, sweeps in its course round numerous islands, some of them inhabited by banditti, and others by hippopotami. Its banks are every where clothed with a broad belt of thorn, willow, and black bark trees, alive with the notes of birds, whilst the strangely shaped hills which so frequently enclose the river, form the most exciting scenes, from their wildness and air of romance, that can possibly be conceived.

It is difficult to speak of the Gariep otherwise than in the most enthusiastic terms. Besides its beautiful African features, its utility is very great. To the wandering tribes dwelling near it affords an unfailing refuge in seasons of drought and famine. Rich grass is always found under the trees. Birds are numerous, and there is plenty of fish, though the natives seldom trouble themselves to draw any supply of food from the waters of the river.

That there will be white men sojourning on the banks of the Orange river at no distant day, I have little doubt; for I found, at convenient distances from the river, great store of valuable iron and copper ores, for which there is always a great and increasing demand in Europe. Their accessibility is their great recommendation here, also their being placed in a dry and healthy climate, and amongst tribes who can easily be conciliated with small presents, and who might even be tempted to assist in working the mines.

But there may be even the more precious metals found along the course of the Gariep; the desolate and dreary regions near the river may yet be found to teem with subterranean wealth of gold and silver, to compensate for their forbidding exterior. But besides the ores we know to exist near the river, the timber is very valuable. I found abundance of ebony, the *Royena decidua*, the bark of which is so well adapted for tanning, and various thorn trees, which would be well adapted for building. Gum might also be collected in any quantity from the acacias along the river, sufficient, I should think, for the consumption of England; of shell lime there is plenty on the coast; and bees' wax also could be procured in great abundance along the river. A Namaqua, who had a waggon, assured me, and I have no reason to doubt him, that on a honey hunt he filled his waggon with skin sacks of honey above the side planks in two or three days.

The Gariep is sometimes a traitor in its sudden rising. A party encamped on the banks may have seen no indications of rain for a long time, when a few clouds may be seen to hang over the mountains in the distant east; and at night the river may come careering down in a mighty flood, spreading over its banks, covering the trees, and carrying destruction to every miserable object exposed to its influence.

Besides trees torn up by the roots and rolling down the flood, sick or wounded hippopotami are sometimes borne down from the upper parts towards the mouth; these, occasionally before they reach the sea are fixed in the trees, and on the subsiding of the waters, they remain (in Dutch phrase) " spurtelen," or kicking among the branches. It must be rather curious to see such monsters aloft; but the natives lose no time in stupid wonder, but quickly dispatch them with their javelins, and make merry over the rich spek, or fat under the skin.

The hippopotami, " whose bones are as strong pieces of brass, and whose bones are like bars of iron," in the upper parts of the Gariep, remain during the day in deep parts of the river, commonly known by the name of sea cow holes (zee koe gatten,) and issue out to feed on grass and foliage at night. A swimmer is in some danger on entering one of the sea cow holes, from the immense mouths and teeth, " terrible roundabout," of the monsters which inhabit them. And when the mighty river horse lifts its head to breathe, the nostrils, small ears and eyes, are in one line, and on a level with the surface of the water, and therefore there is not a large object presented to the marksman. The natives make pit falls for the hippopotami on their paths along the banks.

That there are alligators in the Orange river

I have little doubt, for the dread leviathan,
"whose scales are his pride, shut up together
like a close seal, who maketh the deep to boil
like a pot, and a path to shine after him; before
whom iron is as straw, and brass as rotten wood,"
occurs in the rivers under the same parallel of
latitude on the east coast; and though I am not
aware that any white man has seen an Orange
river alligator, the natives on the banks told me
that at night they heard occasionally cries from
the river like those of children, such as I heard
on the banks of the Mississippi; and that a short
time before I visited the Orange river, a cow
calved on the banks near the mouth, and that a
creature crawled out of the water and devoured
the calf.

There were about one hundred people lying
at Aris, with flocks and herds; they were very
friendly disposed; and Paul Lynx, the chief of
the Orange river, a strapping fellow, with one
eye, and a peculiar savage look, came with two
or three of his counsellors, and laid this case
before me for decision, and on which we sat in
solemn conclave in a mat hut.

That for ten or twelve years Paul Lynx's people had caught the seals on the island before mentioned, had preserved their flesh, on which three hundred had annually subsisted, and had sold their skins. That lately, a white man, anxious to acquire possession of the Seal Island, though it was many miles beyond the border, had actually memorialized the Governor for it, and had shown them a paper which he said was the Governor's answer to his memorial, granting him the Seal Island. He had asked Paul to put his mark to a paper, giving up the Seal Island, or allowing this trader alone to obtain the skins at his own price. The Namaquas then asked me if the Governor had any power to give away their Seal Island; and if I thought he had done so. I said he certainly had no power to grant to any one an island which was at least forty miles beyond the border; and that the paper which had been shown them must be a forgery (which it was), and that they might rest assured that no Governor of the Cape would attempt to annoy them, or deprive them of their property. They then said, " We shall shoot the white man if he attempts to catch seals on our island." I told them on no account to use any violence; but that if any white man (besides their friends Eddington and Kennedy) belonging to the colony attempted to interfere with them, they ought to inform their missionary, Mr. Schmelen, and that he would lay the matter before the Governor of the Cape, and thus obtain justice for them, and the punishment of the intruders.

A little below Aris the Orange river was about five hundred yards wide, and I waded across it without difficulty, to Great Namaqua land.

Having obtained a span of oxen from the people of Aris, we inspanned, and proceeded among sand hills down the left bank of the river. After travelling about twelve miles, and seeing the mighty bones of a river horse on the road, which had been killed and eaten here twenty years before, but not before it had destroyed a hunter; we came to the end of the trees, where the river was seven hundred yards across; and as we approached the mouth we saw several green islands on which troops of horses belonging to the Namaquas were feeding.

We outspanned at three miles from the mouth, at a post or beacon on which " De Graaff, Land-drost, 1809," was carved, and notwithstanding a slight shower of rain, we slept soundly among the bushes.   In the morning we were awoke by the cries of innumerable flocks of wild fowl. We soon jumped up and stole down with guns to the bank, to get a shot at the wild geese, which were leading their young to the water, over the soft mud, whilst flamingoes, with snow-white bodies and red beaks and legs, and rosy blush on the wings, stretched their long necks over head, and like winged rods flew round us. Wild ducks with reddish breasts, rose noisely from the river, which now expanded into a lake of four miles in breadth.   Flocks of pin-tailed grouse flew swiftly about us, with their lively note, " tsué, tsué;" sand-pipers ran along the wet sand and mud, and, in short, the quantities of wild fowl we saw here were immense, the air was darkened with them, and the shore con-stantly resounded with their cries—" *littora re-sonant strepitu.*"

Hares I found in plenty at the Orange river

mouth; there is also the large elandbok to be found here; and an immense snake is occasionally seen whose trace on the sand is a foot broad. The natives say, that when coiled up, the circumference of this snake is equal to that of a waggon wheel; and when it visits the Orange river mouth it is a sign of a good season for rain.

We mounted horses and rode along the lake to the mouth; the country was everywhere flat, and without landmarks; and as the mouth of the river overlaps as it were, it is very difficult to hit it from seaward. The Twin, or Buchu, Mountains, at Cape Voltas, a few miles to the south, is the best point to make for, previous to searching for the mouth of the Gariep.

Vast quantities of excellent dry timber for firing lay about, brought down by the floods of the river. At last we saw a line of breakers assailing the sandy beach with hollow roar, and stretching right across the mouth, which was merely an opening of about one hundred and seventy yards, between two points of sands, on which sat a line of penguins and gulls:—outside lay the ocean;—

> Beautiful, sublime, and glorious,
>   Mild, majestic, foaming, free!
> Over time itself victorious—
>   Image of eternity!
>
> Such art thou—stupendous Ocean!
>   But if overwhelmed by thee,
> Can we think, without emotion,
>   What must thy Creator be?

I saw no rocks or dangers here, nor did I hear of any rocks from the natives, in or about the mouth.   There is probably a shoal of sand outside; but, with care, it seemed that the mouth of the river could be entered by a schooner. I never heard that any soundings had been taken at the mouth of the Orange by the African surveying vessels : and the increasing wants of the Cape Colony demand a far more detailed survey of the south, south-eastern, and south-western coast of Africa, than any that exists at present.

From there being so much wild fowl at the mouth of the Orange, plenty of excellent fish; as "springer" and "harder," for the seine, and abundance of grass and fire-wood; a few industrious families could make a good living here, taking care to "pitch their tents" out of reach of the line of inundation.

I asked for bays about the mouth, and was told that a short distance south of it (say two miles) there was one. I accordingly rode there, and found a boat bay, or inlet, into the rocky line of coast, of about five hundred yards in length and two hundred across, in which small craft might securely anchor. I was also told that there is another bay near Cape Voltas; but which I had not an opportunity of visiting.

We rode back to our waggon across a country composed of sand and scattered bushes, and in which numerous mole-holes rendered riding dangerous. A feeling of heart-sinking took possession of one in traversing these hot, flat, and lonely wastes, on which there was no object to interest one, and no sound struck the ear, except the distant bellow of the breakers, and the clamour of the wild-fowl at the debouchure of the Great River.

After some more skirmishing with the geese, for our pot, we inspanned, and again tracked over the twenty miles to Aris.

I now made, by means of an old Bastaard— William Joseph, the interesting discovery of a large mass of copper, which exists about sixty miles E. S. E. of Aris, and about twenty miles from the south bank of the Orange river. This copper is quite accessible, and it might be either smelted on the spot, with Orange river wood; of which, as I said before, there is plenty; or the ore might be floated down to the mouth on rafts, which rafts might then be sawn up, and sent to the Cape, where wood is always in great demand for building; or the ore might be transported in waggons to the sea, and shipped from Cape Voltas, or the mouth of the Great River.

A son of William Joseph was hunting in 1834, when he came to a spot abounding in green stones, and following the direction of these, he found masses of green rock. He brought home pieces of these, and on melting them a blue flame arose, and bright copper ran in the fire, from which bullets were cast.

I brought away a quantity of this ore from the river, which was assayed by Sir John Herschel, at the Cape, and from a picked specimen, sixty-five per cent. of metal was the return; another specimen, taken at random from the others, yielded twenty-eight per cent. in London. Now the richest of the South American mines yield only twenty-five. In consequence of the discovery of this *accessible* Orange river copper (and there is also I know rich copper one day's journey north-east of Keerom, within the colony), several men of business in London have communicated with me regarding the establishment of an Orange river copper company.

The natives, as I mentioned before, are friendly disposed. None occupy the ground where the Orange river copper is, and if white strangers were kept under proper restraint and control, the natives would be pleased to see them among them, for the sake of the articles of European manufacture which would be introduced among them. The natives might even be induced to assist in working the mine. There is no sickness upon the Orange river, and the heat is troublesome only during our European winter months.

By reference to the map, it will be seen that I subsequently found iron not far off. There

are also indications of coal near the iron, and as
the site of these valuable ores of iron and copper
is far beyond the colony, no expense would be
incurred in purchasing a right to work them,
and it is to be hoped that they will before long
be turned to good account.

The Aris people brought me a fine leopard,
which they had just killed with a string-gun, set
among bushes.   The leopard had been troublesome among the horses.   The manner in which
the gun is set is simply thus:—two short stakes
are driven into the ground near the leopard's
haunt, at the distance of three feet from each
other; the ramrod is taken out of the loaded gun;
a short stick is tied across the stock in rear of
the lock; from one end of the stick a string communicates with the trigger, from the other end a
long string leads past the cock and through the
ramrod guards; and the gun is then fixed to
the stakes at a height of fifteen inches above
the ground, to take a leopard in the breast; the
other end of the long string communicates with
a piece of meat, and the gun is carefully concealed with the bushes. Some skill is required to

fix the muzzle of the gun at the proper height for wild animals, from a lion to a jackal, but the natives are expert, and the charge seldom fails to take effect.

Whilst the river was rapidly filling, we left Aris, travelled south again, and saw many ostriches. I left the waggon, and went across the country on foot in pursuit of a pair of these gigantic birds. I thought I could manage to get within shot of them by creeping and running among the bushes, but I found out, after a hard run of half an hour, why long necks were given to ostriches. They looked over the bushes every where, and with their black bodies and white tails, and wings outspread, their pillar-like legs and the wind soon carried them far out of my reach.

According to native testimony, the male ostrich sits on the nest (which is merely a hollow space scooped out in the sand) during the night, the better to defend the eggs from jackals and other nocturnal plunderers; towards morning he *brommels* or utters a grumbling sound, for the female to come and take his place; she sits on the eggs during the cool of the morning and evening. In the middle of the day, the pair, leaving the eggs in charge of the sun, and " forgetting that the foot may crush them, or the wild beast break them," employ themselves in feeding off the tops of bushes in the plain near their nest.

Looking aloft at this time of day, a white Egyptian vulture may be seen soaring in mid air, with a large stone between his talons. Having carefully surveyed the ground below him, he suddenly lets fall the stone, and then follows it in rapid descent. Let the hunter run to the spot, and he will find a nest of probably a score of eggs (each equal in size to twenty-four hen's eggs), some of them broken by the vulture.

The jackal is said to roll the eggs together to break them, whilst the hyena pushes them off with its nose to bury them at a distance.

On the 5th of November we were again at Ukribib, thermometer 87° at noon, wind south, and at the Kowsie on the 6th the heat was 103°, with a parching wind from the north. We shot a couple of steenboks, but had great difficulty in getting water in the Kowsie at this time. We dug with our fingers in the sand, and got at last, by creeping on our hands and knees into a hole, some brackish and fetid water, to moisten our meat, with which we had neither bread nor salt.

A tall and beautiful secretary bird, with its blueish plumage, its " black breeches and grey stockings," and quills stuck behind its ear, marched along fearlessly and unharmed near the waggon, on the look out for snakes, of which it is the mortal enemy. It pounces on them with its strong feet, and shields itself with its wing, when the reptile, with swollen head, red eyes, and horrid mouth, erects itself in vain to intimidate the bird. A tame secretary, on seeing a rope drawn along the ground, mistaking it for a snake, will dash at it, and trample upon it fiercely.

Again we reached Komakas; and inspanning a light horse cart, I drove off with 'Nolus Vanzyl to Keerom, by a very rough mountain road.

A bull gemsbok (which with its long and straight horns is a match for a lion, single handed) had been lately about the kloofs at Keerom; 'Nolus sent out his son to watch and shoot it; he did so, after some trouble, and on cutting it up, no less than thirteen balls were found in it, besides the marks of others which had gone through. The gemsbok is well known to the natives to be very tenacious of life, whereas the eland is easily killed, and soon dies even if it breaks its leg only.

The height of a bull gemsbok is about three feet and a half at the shoulder; the horns two feet and a half; the tail is a black switch. The white face is crossed with two bands of black; the general colour of the body is iron-grey, which is separated from the white belly by a black band. From the marked contrast of colours on the gemsbok, and its formidable horns, its appearance is very wild and warlike, and only young lions venture to attack it.

I rode by Wolfpoort and the Cardow to Lily Fountain, where all my people were well, and anxious to proceed; and here I now found the Rev. Mr. Edwards, with his wife and family, returned from the Cape, and the chief Abram,

of the Bondelzwart (bundle of blacks) Namaquas, with three men, waiting to see me.

Abram is one of the ugliest men that can be
conceived. His figure is tall and good, but his
face is most disagreeable to look on, with a flattened nose, wide mouth, hollow cheeks, high
cheek bones, and narrow eyes. He was dressed
in a claret coloured jacket and leather trousers.

I gave Abram a present of a green cloak and
medal (with the cipher of the king on it) from
government, and a pipe and some small articles
from myself; and having thus endeavoured to
please him, I asked him what he thought of a
journey towards the Damara country. He said
he was very much afraid of it (and it was evident
by his bearing that he was a coward), that the
Damaras were very wild, that he had no influence
beyond seven days journey north of the Warm
Bath, that in his own country he would do what
he could for me, but that he was at variance with
the Orlam Namaquas beyond him.

He was also alarmed about the Great river filling
again, was very anxious to go back to his kraal,
and on my asking him if he thought I could get

pack oxen to purchase among his people; he said, he did not think I could. This rather surprised me, and determined me to pick up some if I could about the Kamiesberg:—all I could get, were six from Mr. Archer the trader.

Abram promised to come with his swimmers to assist us over the Great river, and then took leave. Till the field-cornet and the twelve Boors were ready to escort me, by order of the Governor, over the river, (to produce a good impression on the natives), I spent a few agreeable days on the mountain.

I said before that many of the old farmers consider the coloured classes in so inferior a light, that they think them quite unworthy of worshipping the Deity in common with themselves, saying that the Bible is only for white men. I now saw another striking proof of this ignorant feeling. Whilst I was hearing the mission school examined (and certainly the proficiency of the children in reading and understanding the scriptures was very great) I remarked an old woman with spectacles sitting beside two or three other women, and attentively reading the sacred volume. I asked who she was, and the missionary told her to answer me. She informed me that she had lived with a farmer who would not allow his people to hear the family worship which he occasionally held, but that she used to listen at the window and behind the door, and thus learnt something by stealth; could repeat the Lord's prayer, and could sing a hymn or two, still she wanted something more, and she searched for it in vain; but after a time she overheard an old man praying in a wood, and thus she was first instructed in her religious duties. She came to the missionaries, learned to read when she was advanced in years, and now seemed to take a wonderful delight in religious exercises.

Mr. Edwards had got some good Merino rams on the mountain, and a flock of four hundred cross-bred Merino sheep, and he was desirous of instructing the people in making cloth to cover themselves in winter (instead of wretched skins), and to manufacture hats and shoes. An English schoolmaster, as an assistant to the missionary, seemed also to be particularly required on the mountain. It is now high time to leave off instructing the native children in South Africa through the medium of the Dutch language. I do not know any mission station where English reading and speaking are exclusively taught, as they ought to be, seeing that we have now held the Cape Colony since 1806, and are likely, I trust, to hold it in undisputed possession for many ages to come.

A Namaqua came on the mountain, who said that he had lately been at the kraal of Stuurman the robber chief, for whom a great reward had been offered by government. Stuurman with a band of Corannas and Boschmans had, a few years ago, made an inroad into the north eastern part of the colony, and had destroyed several Dutch families, men, women and children, and had plundered and burnt their houses. A commando of armed farmers had gone out against him, but had failed to take him, and now his haunt was discovered for the first time. It was on an island several days journey up the Orange river, well fenced round with trees and

bushes, but through which our Namaqua in-
formant said he could guide a party. Of course
I lost no time in communicating with government on this subject, and steps will, doubtless,
again be taken for the murderer's apprehension.

The field-cornet and his Boors were very
averse to what they considered the very dangerous undertaking of accompanying me across
the Great river. As to taking Stuurman with
their assistance, seemed impossible, though some
farmers could be depended on for courage and
daring, particularly old 'Nolus and his brother
Piet Vanzyl; but they had not been ordered
out on this occasion. None of my party had
ever been beyond the frontier before; and they
now dreamt and talked of nothing but lions and
Boschmans, and I have no doubt that they and
their "poor wives" gave me many a curse. Still
go they must; and though the field-cornet (Agganbag), came and wished to delay the journey,
saying that the warning was " Zeer kort, Mynheer," (very short,) I would not yield, and said
that we must move from the mountain on the
16th positively.

G 3

The vrows and meisjes were accordingly at work day and night at the farm-houses below the mountain, preparing biltong or dried meat, and baking bread for their men; they mingled tears with their handywork, thinking that their devoted people would with difficulty escape with their lives from the perils of the Orange river wilderness. Nothing was talked of in the country but " de schrikkelijk reis," the terrible journey!

CHAPTER VI.

Leave Lily Fountain with a Caravan — Weeping and Wailing
— Silver Fountain — Murder of the Missionary Trelfall
described — Mr. Schmelen — The Copper Berg — Foun-
tains failing — The Corn Mill — Bezondermeid — The
Rev. Mr. Wemer and his Privations — Three great Dan-
gers in South African Travelling — Hills of Pandemo-
nium — Ostrich Chase — Henkrees — Distances deceitful
— The Silurus Gariepinus — Burchell the Traveller —
Old Balli — The Blood Sickness — Tales of the Interior
— A Dance — A Hard Night's Work — Ford of the
Bustard, Orange River — Rev. Mr. Jackson and Abram
the Chief — Why the latter desired to have a Missionary—
A Lion Hunt — Sporting on the Orange River — Story
of a Baboon — Namaquas dislike Fish — A Sea Fisher —
My Horses — Notions on Naming — " Fight it out " —
Honey Beer — The Escort dismissed — A False Alarm
— Reach the Warm Bath.

My party, on leaving Lily Fountain on the
16th of November, formed a small caravan :
there were, besides my own men, the field-cornet
and twelve strapping Boors, each riding a long-
tailed horse, leading another, and armed with
elephant guns carrying four balls to the pound :
there were also two waggons, fifty oxen, fifty
sheep, and half a dozen dogs.

We travelled north, and descended by a narrow and rocky road, which resounded like a smith's shop, to Floris Coetzer's place (four hours with an ox-waggon), on a branch of the Buffalo River. Here in the evening the thermometer was at 53°, the wind was westerly, and the cold most intense; there was no possibility of keeping oneself warm.

It was very laughable to see the view which the Boors and their wives and daughters took of the journey to the Orange River. Some of the women came to take leave of my escort on the first day's march. The vrows cried and roared, and the Boors joined them with tears running down their cheeks. Truly it ought to have been very affecting, for it seemed as if the bulky bodies of the escort were assuredly destined to feed the hungry lions of Great Namaqua Land, or to be butts for Boschmans' darts.

Our next outspan was at Silver Fountain (nine hours). To reach it we passed first through rocky and *shaky* passes, with walls of rock on each side, and where I thought the waggons would have been dashed to pieces every instant, and then through a fine open country, with scattered hills and Euphorbia-covered plains. The farm·house at Silver Fountain is on a plain, with a garden of fig and peach trees, through which water is led, and near it are some rocks, below which is the grave of the murderer of the missionary Trelfall.

The Rev. Mr. Trelfall was a young man of great zeal in the sacred cause of converting the heathen. He had been some time in the pestilential climate of Delagoa Bay, and on his return to the Cape to recover from a fever attended with delirium, he resolved to try the west coast of South Africa, and to penetrate to the Damaras. His undertaking was a rash one; for he took no precautions for defence, and no one can travel far in Africa trusting only to the good will of the natives, " whose tender mercies are cruel."

Mr. Trelfall crossed the Great River, and reached the Warm Bath in 1826, with two Hottentots (Jacob and Jan), and three pack-oxen. Here he got the guide, Naugabib, who murdered him. The murderer confessed as follows.

That the chief at the Bath asked the missionary for powder, who being unable to furnish a supply, was desired to leave the kraal, which he did; but being forced to come back for water, the chief told Naugabib to murder the missionary when he got some distance on the road, and to call in assistance if required. Naugabib said that, being afraid of the chief (who barbarously cuts the sinews of the necks of those who offend him), promised to obey his orders. They again left the Bath, and travelled north-west towards the Fish River, and falling in with a hunter armed with a gun, Naugabib persuaded him to join the party, though the missionary objected to it, owing to the difficulty of finding food.

They had " packed off" near the Fish River, and on Naugabib being refused some tobacco he wanted from the missionary, he got in a great passion, and threw back the trowsers and handkerchief he had got for guiding the party. The missionary then pacified him, and gave him some tinder-boxes and beads, to go to a Boschman kraal near, and endeavour to get some food. The guide went to the kraal, and told the Boschmans what the orders of the chief were regarding the disposal of the missionary; the guide slept at the kraal one night, and next day having arranged that five Boschmans should join him in the evening to carry his purpose into effect, he brought a goat with him to Mr. Trelfall.

In the evening, the Boschmans, armed with bows, arrows, and javelins, came and sat by the fire; and after the missionary had sung a hymn and prayed, he and his two Hottentots lay down to sleep in the sand. When all was quiet, Naugabib went to the Boschmans and said, "Now is the time." The guide, the hunter, and the Boschmans accordingly surrounded the sleepers quietly. The Hottentots were first assailed with arrows and stones, and Jacob's backbone was shot through with the gun, on which the missionary awoke, and asked if there were lions near, and getting no answer from his people, he rolled himself up in his kaross and again went to sleep. The hunter now said it would be better to leave the white man to die of hunger and thirst in the wilderness, which he would certainly do, as he could not find food by himself; but the guide said, " No ! he must be killed like the rest."

. Accordingly, Naugabib pulled the kaross off Mr. Trelfall, who getting up and seeing the murderers round him, immediately understood the desperate circumstances he was in, and putting his hand to his breast as if to search for a weapon, he fled towards a neighbouring bush. Naugabib and the others at first hesitated to follow him, thinking he was armed, but seeing no weapon, arrows were discharged at him, and then an assegaye was thrown. The devoted missionary next fled towards the packs, and knelt down and prayed, (doubtless for deliverance from the bitter cup of death which he was then tasting); but, poor man ! his fate was sealed. The infernal Naugabib knocked him down with the blow of a stone on the temple, and then ordered the Boschmans to destroy him with their assagayes, and also to finish Jacob, who was still groaning in the agonies of death. They did so; and then the baggage was plundered, for which the murders had been committed. Mr. Trelfall and Jacob were also stripped; but Jan's clothes were too full of blood to be of any use: the three bodies were left to the wolf and the vulture.

Thus perished Mr. Trelfall, the victim of imprudent confidence among savages. He had pistols, but he did not load them, nor had he one about him to shew when he was assailed, or he might have frightened his cowardly murderers.

Naugabib, after taking the pack-oxen, and giving the hunter and the Boschmans some knives and tinder-boxes, went to his own kraal, and did not return to the Warm Bath, and he and the hunter were captured some months after by the Orlam Namaquas. Naugabib was shot at Silver Fountain by the people of Witboy, the chief of Pella, in presence of Mr. Ryneveld, of Clanwilliam, and of Abram of the Bath. The hunter was confined, and was not shot, as he was only an instrument in the hands of the monster Naugabib.

Mr. Schmelen came to meet me at Silver

Fountain's, having escorted the Rhenish missionaries part of the way to Pella. He generously brought me a supply of flour, which was very acceptable, and he showed me the graves of the wives of two German missionaries, (who had wandered thus far with their husbands many years before), Mrs. Albrecht and Mrs. Sass.

> " Thus at the shut of even, the weary bird
> Leaves the wide air, and in some lonely brake
> Cowers down, and dozes till the dawning day,
> Then claps his well-fledged wings and bears away."

On the 18th we rode on between hills as usual, and by a narrow pass among them, on the sides of which were scattered the strange koker boom, and after seven hours arrived at an outspan place, where there was water under the Copper Berg, a mountain on which there are frequent indications of the valuable metal from which it takes its name.

The fountain at the Copper Berg had much decreased within the last thirty years. Formerly a thousand head of cattle could at most times have drank at it; now there was scarcely water enough for a span of twelve. The old people said that much less rain had fallen within the latter years—that there was no sea rains now as there used to be, only thunder storms from the east; but they hoped that the following years would take a turn for the better. An old Bastaard, who had lived in the colony in his youth, said, that long ago, he remembered that the shadow of the sun, in the longest days, fell at his feet; now it fell some distance from them, at mid day. " This," said he, " may in some way be connected with the failing of the fountains."

Three fat women, grinding corn at the mat huts of a Bastaard, Lang Cobus Cloete, had a curious appearance. The millstones were fixed close to the ground, in an upright frame, with a long handle; the women, stripped to their waist, stood up to their work, and their attitudes and figures, and flapping skins, whilst turning round the handle, had a very grotesque look.

Five hours brought us to Koe Kus, on a river where there was plenty of grass and wood, and eight hours brought us to Bezondermeid (strange woman). This was the station of the old labourer among the heathen, the Rev. Mr. Wemer, a German, under the London Missionary Society.

Nothing can be conceived more desolate and forbidding than the appearance of the country about Bezondermeid, at this the hot season of the year. Some black and bare hills bounded sandy and bare plains, on which, beside the dry bed of a stream (in which there was one hole for water), stood three or four mat huts and a waggon. This was the picture of the station.

Mr. Wemer, living like his flock in a mat hut, was seventy-four years of age, and during a great part of his life had moved about among three hundred of the natives south of the Great River, he and Mr. Schmelen being the only two missionaries between Lily Fountain and the Great River. Mr. Wemer had a very tidy native wife, by whom he had several children; one was in arms when I was at Bezondermeid. Ye dwellers in cities, nourished in abundance, think of the life this old man was leading! and yet he seemed cheerful and contented, particularly when I gave him a supply of tobacco, of which he was in want to replenish his meerçhaum pipe. But for a year he had not tasted bread, and had been out of salt for six months: till my waggon came up, he regaled Mr. Schmelen and myself on boiled salt beef and Bush tea.

Could not some portion of the funds of the London Missionary Society be applied to pensioning off such aged and faithful labourers as Mr. Wemer—so that in their latter days, after years of privation and hardship, they might partake of, at least, the common necessaries of life? I am sure it is unknown to the directors of the society what some of their old missionaries suffer.

Mr. Schmelen now took leave of us to return home; he seemed considerably affected at parting, for he knew what we had to go through beyond the Gariep, and doubtless he thought that it was twenty chances to one that we should ever be south of the Great River again. My people and myself were highly pleased with the good humour and kindness of heart of this worthy old man, and we parted from him with great regret, though he offered to go further with me if I thought he could be of use among the natives; but I could not think of taking him out of his road. " I trusted to Providence and a good look out," to pass in safety through the three great dangers in South African travelling, savage men, wild beasts, and want of water.

" Chance will not do the work : chance sends the breeze ;
But if the pilot slumber at the helm,
The very wind that wafts us towards the port
May dash us on the shelves. The steersman's place is vigilance,
Blow it or rough or smooth."

Travelling N.E. and N.N.E. we reached a small oasis in the dry and stony wilderness at Eerebies, where in a small hollow in the plain, beside a dam of water, stood two mat huts in a garden, containing tobacco, melons, pumpkins, and a few heads of wheat. At a corner of the garden, bushes were put up to conceal a person, who could thus shoot the pintailed grouse, which frequented the water in large coveys.

The country now got worse every mile : nothing was seen on the sandy plains but a few stunted bushes, and the hills, of a few hundred feet elevation, were black and red, as if burning, whilst fragments had crumbled off them, gnawed by the devouring tooth of time, and lay at their base. " A dassé could not live here," said a

Boor, " only the klip salamander," (or rock lizard).

The appearance of nature here was as that of a land accursed. We seemed to be on the shores of the Dead Sea, where the cities of the plain had sunk under the fiery wave, and where desolation for ever reigns around, to mark an awful judgment; or it appeared as if the glowing hills of Pandemonium had been raised from their dreadful depths to sully the face of the fair earth with their most forbidding aspect.

Thermometer in the shade at mid-day 98°.

Two ostriches appeared before us, and as it was hot enough to pursue them, a chase took place after them, with horses; but though they are most likely to be come up with in the heat of the day, on this occasion the chase was in vain: they fled before the wind, " scorning the horse and his rider."

Seven hours from Eerebies brought us to Henkrees, five or six miles only from the Orange river, the bed of which we saw inclosed with fantastic and peaked hills. My white attendants being very anxious to see the Great River, asked leave to go and fish there, as the distance seemed to be nothing; but appearances are very deceitful in the clear atmosphere of South Africa: hills which appear to be only three miles off are probably seven; and so it proved in the present case, for when my people came back in the evening with two or three large moekul or flat heads, they were knocked up with the distance, the heat, and the sand.

The flat head, or Silurus Gariepinus, (so named by my friend Mr. Burchell, one of the most painstaking, industrious, and intelligent of travellers, and whose great work on the central regions of South Africa cannot be sufficiently admired for its minute accuracy), the flat head, I say, is about three feet long, of a dark green colour above, and white below; the skin is without scales; the head is compressed, the mouth bearded with six long rays, and the eyes small and yellow. The taste of this, the commonest of the Great River fish, and of the streams which flow into it, resembles that of the eel.

A respectable old Bastaard lived at Henkrees, Balli by name; he was the owner of a thousand head of cattle, of many horses and sheep. Unlike the generality of the people of the land, he and his sons were very lusty. The united families formed a small community of about thirty persons, in mat huts. Balli was suffering under the disease called blood sickness. Sheep occasionally get this, which is a corrupted state of the blood; the natives then kill them, and, strange to say, eat them. The disease is very often communicated to those partaking of the impure flesh; and it appears in angry sores on the arms and on other parts of the body. It is very difficult of cure. A person starving may be excused eating bad meat, but not one, like old Balli, possessing flocks and herds.

I sat some time with him in his hut, and he told me of monstrous snakes which he had seen in the land, whose presence was indicated among long grass by their smell, which was most offensive to cattle, and whose bulk, when coiled up, occupied a space as large as an after waggon wheel. He also told me of great quantities of copper far up the Great Fish River, and of hills from which maleable iron could be cut out. And he talked of the wars between the Namaquas and the negro Damaras.

In the evening, the violin, tambourine, and castanets, set the party dancing. Some of the lusty Boors tried to cut capers, and the Namaquas performed a sort of reel, the men showing wonderful activity, springing into the air and striking their soles together, and the women sailing about in their large karosses. There was no impropriety shewn in this dance; and as the people seemed fond of it, I promoted it to keep them in good humour, by means of which I hoped to "progress" more smoothly. I myself, of course, refrained from dancing, though very fond of it.

Balli was very civil, gave me the loan of fresh oxen to help us to the ford, (which was a day's journey higher up the river,) and an Orange river dog, small, yellow, and with a black muzzle and curled tail. On the evening of the 23rd November, we set off, and traversed S.S.E. to avoid the hills, which run parallel with the river, and then E. We had a severe night of it: we were twelve hours inspanned, toiling through heavy sand. Sometimes the waggons stuck fast, when a party was called to the wheels to assist the poor bullocks. There was incessant shouting and cracking of whips, and much fatigue. About four in the morning, when the ground got harder, and we began to descend to the river, I rode on to look at the ford, reached the Great River through an opening in its great wall, and saw it half full; that is, two hundred and fifty yards of the five hundred of its bed were covered with water. I saw a considerable party of men on the opposite side, and, overcome with sleep, I donned my pilot coat, and with my saddle for a pillow, I slept soundly under a bush till the waggons came up.

The ford of the Karahas, or Bustard, is not shallower than other parts of the river near, but it affords greater facilities for entering, and emerging from the river. We had no need on this occasion for the assistance of the swimmers, for such we found the party on the north bank to be; and fording the river up to our middles, the waggons were dragged through without injury to the stores, and we found ourselves in the midst of the chief Abram and sixteen men, accompanied by the Rev. Mr. Jackson, from the Warm Bath, the only missionary, or white man, in Great Namaqua land.

Mr. Jackson belonged to the Wesleyan society. He and a Mr. Cook had recently arrived in South Africa; and on the former proceeding with Mr. Edwards from Lily Fountain to look for a situation for a new station, they fell in with Abram, who said he wanted to have a missionary with him. Accordingly, Mr. Cook established himself with Abram's tribe, and being now absent at the Cape, Mr. Jackson was occupying his place.

People ignorant of the sulky and selfish character of Abram, would give him great credit for the desire he expressed to have a missionary, thinking that he was most anxious for the improvement of himself and his people, whereas he had no such desire; his principal object really being to increase his influence in Namaqua land by the presence of a missionary at his kraal, who would also enable him to get a supply of gunpowder from Government, and would furnish

him with cutlery, cloth, and other articles which
he and his people coveted. Still it was desirable
that a missionary should be placed with Abram,
whatever the chief's motives were for wishing
to have one with him, for his tribe is large;
and though the people of it bear a very indifferent character in the land, it was desirable to
endeavour to improve them by missionary influence.

A lion had just been killed by Abram and his
people, and Mr. Jackson had been present at the
death The king of the wilderness had destroyed
two horses in the field; and the people, armed
with guns, set out to spoor or track him to his
lair. They roused him, and he trotted clumsily
up a ravine, and then lay down under a tree;
fifty or sixty of the people then fired at him
without effect, on which he rose to the charge—
his mane erected, his tail lashing the ground like
a flail, and his throat roaring defiance. A party
scrambled up the rocks behind him, when one
lucky ball from the chief struck him above
the eye, and levelled him; thirty more shots
were then thrown away by the people behind him
on his dead carcase.

I went to shoot wild ducks in the river, and my people set about fishing. It is not altogether safe to traverse alone the banks of the Orange. Besides lurking Boschmans, with their poisoned arrows, lions are to be met with, panthers, and, above all, *baboons* are to be dreaded. The large dog-faced baboon, five feet in height, very strong, and covered with black hair, will not hesitate to attack a man if he find him alone, to attempt violence to a female, or to carry off a child.

After my man Robert had left the South-Seaman (to which he belonged) at Angra Piquena Bay, he staid about the Orange river for some time; and one day, while fishing alone under the trees, he was diverted by the gambols of some young baboons on the opposite cliffs, when suddenly he heard a loud " quah " behind him; and looking round, he saw a great baboon close to him. Robert had no weapon to defend himself with. The hairy monster cried "quah" again; when a number of other baboons were seen rapidly descending a neighbouring hill. There was no time to be lost—Robert snatched up a branch which he found at hand, and when the baboon was closing with him, and shewing his horrid teeth, with the intention of biting him to death in the neck, Robert struck desperately at his head; the baboon put up his left arm, and received the blow on it, and immediately wrested the stick out of Robert's hands, though he was a strong sailor. Flight was now Robert's only chance, and he took to his heels as fast as he could, followed by the baboon, who, though partly crippled by the blow, still "quah, quahed" after him, till Robert gained the open country, and the Namaqua encampment from which he had come, appeared;—the baboon then gave up the pursuit.

The baboons live on scorpions, lizards, bulbs, and gum, with which last their paws are usually smeared; they sleep in holes of the rocks; are seen in troops of a dozen or more, headed by a large male, the females bringing up the rear, with their young clinging to their backs. They seat themselves on the ridges in the morning and evening, and carrying their tails as if broken in the middle. Before mid day they descend nimbly over the open slopes to the cover of trees along the river. With their disagreeable " quah," of different cadences, they make the lonely banks of the Gariep to re-echo.

I had some sport with my gun, then bathed in the river, and went to see how my people had succeeded with the hook. They had caught not only moekul, but large carp in considerable quantities;—these last were eighteen inches or two feet long, of a greenish colour on the back, and with yellow sides. I gave some fish to the Namaquas, but they declined eating them, saying that they might be poisonous. On my setting them the example, and giving them plenty of sheep's tail fat to eat with the fish, they ventured to taste them. Strange ! that people who are often short of food, and are compelled to eat gum and roots, should neglect the inexhaustible supply which the Great River and its tributaries afford.

Robert had a good way of collecting fish in one place :—he took the paunch of a sheep, cut a small hole in it, and then dropped it into a still part of the river; in an hour or two, fish, in crowds, surrounded it, watching for what came from the paunch; Robert then cast his hook among the expectants, and landed fish very fast. I having been a fisher in Scottish rivers, was particular in covering my hook with the bait:— "There is no occasion to mind that," said Robert, "that is gentlemen's fancy; we sea fishers clap on bait any how, so"—and he tore off a bait from a piece of flesh with his teeth, and put it carelessly on the hook, and did not the less succeed in taking fish. When he got one in his hand, he gave the back of its head a "scrunch" between his teeth, and threw it down quite dead, saying, "that is the readiest way to settle them, sir."

My bay horse was now so knocked up with the desert through which we had lately passed, that he could hardly move one leg before another. (Certainly there was a contrast for both man and horse between the fare at Government House and what we now got in the wilds.) I left him at the river to come on slowly after us, and was obliged to buy a strong horse, called England, from the field-cornet. My "grey Night" held out and got better into the way of subsisting on the tops of bushes than the poor bay did.

I named the highest summit opposite the ford of the Karakas, Mount Maconochie, after my friend Captain Maconochie, R.N., the first secretary of the Royal Geographical Society; but, during the journey, I named no mountain or hill which had a native name. Nothing can be worse than giving European names, when there are already native designations.

On the 25th we left the Orange river, and travelling N. E. for eight hours, between hills where there was lion and zebra spoor, we reached Sand Fountain. At mid day on the 26th, the thermometer was at 97° under the fly of the waggon, which was contrived so as to form a sort of verandah, with the assistance of lines and boarding pikes. Though this roasting was not very comfortable, I was not much distressed by it, having already learnt how to mitigate it, by abstinence, under East and West Indian suns; my chief anxiety was to make my people bear it with composure, and to keep them in good temper, for the heat made them irritable, and there had been already one or two battles among them when my back was turned. I told them that if they disagreed, they had better strip and fight it out at once behind the bushes; but on no account to allow any disunion among the party to appear before the natives, who might think it worth their while to attempt to take advantage of it, when there would be an end of the expedition.

There were two or three mat huts at Sand Fountain, and the people in them were amusing themselves drinking honey beer, made with honey and water, mixed in a bambus, and fermented by means of a root called " mor," but which I only saw when ground. This honey beer is quite sufficient for the purposes of intoxication. The old head of the Sand Fountain party of Namaquas, sat all day with a bambus of beer between his knees, in one of the huts, quite " hazy" and confused with liquor; and some of the women were little better. After such a state of things, few children " can know their own father."

A few hours brought us to Ahuries Fountain, which was a hole in the bed of a periodical river, closely surrounded with hills. Here I was met by another waggon from the Bath, and I was therefore now able to part company with my escort and the field-cornet's waggon. The Boors most gladly turned their face home again: they had had quite enough of the sandy plains of the Orange river, and of water impure with cattle and wild beasts.

Still the fear of the Gariep rising tormented them, and of being detained on its banks, and exposed to the attacks of lions and of Boschmans, of whom they continually dreamt. Thus, one of them was preparing his dinner, on one occasion, under a tree, when the horns of an ox suddenly appearing among the bushes behind him, he scrambled up the tree as fast as he could, calling out to his comrades that two Boschmans were upon them, armed with guns; but discovering his mistake when he had got up some height, he descended, but not before, a hungry dog had carried off his dinner !

All the Boors got home alive, but killed some of their horses with hard riding—" Post equites sedet atra cura"—and will doubtless talk for the rest of their lives of the journey to " de Groote Rivier."

Leaving the waggons to follow on slowly, on

the 27th of November I rode ahead, with a guide, past Looris Fountain, where were a few Namaquas, also drinking beer, and then I reached a plain on which, beside some rocks, were two stone houses, surrounded with about fifty mat huts: this was the kraal of the chief Abram, and commonly called the Warm Bath.

CHAPTER VII.

Sketch of the Warm or Nisbett's Bath — Ablution becomes
fashionable—Training Pack Oxen—Honey Legs—Springboks — Extent of Knowledge among the Namaquas —
Diseases — Heijé Eibib — Witchcraft — Anecdote — Namaqua Notion about the Sun — Story of the Moon — Summary of Namaqua Customs — Questions put to Old Men
regarding the Occupations of the People — The Choice
of a Chief -- The Missionaries, &c. — A Lion Story —
Sunday at the Bath — How to make a gun shoot straight
—Saturday Night — Excursion to Elliot's Hill — Set out

THE Warm Bath, or Nisbett's Bath, as it is now called (in honour of a Mr. Nisbett, who advanced a considerable sum for this station) is a remarkable place in Namaqua land, as it is the head quarters of one of the most considerable tribes. It was convenient for me to " set up my staff" here on the banks of the 'Hoom for a time, that I might wait for the thunder rains which fall about the beginning of the year, previous to attempting to penetrate further to the north. I got my people therefore comfortably placed in a large shed, whilst I occupied one of the three rooms of Mr. Jackson's house.

The Warm Spring was about five hundred yards from the house, and among rocks. The water continually bubbled up from two or three " eyes," and the heat was of the agreeable temperature for bathing of 105°. A stream ran from the fountain only six inches broad, and an inch and a half deep, yet this served to irrigate a

tobacco and melon garden below, in which the
chief and some of his head men had plots, but
which laboured under the disadvantage of having
a brack soil.

One of the first things I now did was to clear
out, with the assistance of my people, the sand
and stones in the bottom of the bath, and to
make it sufficiently deep for the purposes of
bathing. We set the natives the example of
ablution, and it immediately became fashionable; the old women in particular used to sit in
the tepid water for a great part of the day; and
men and women became exceedingly fond of
clearing themselves from the grease with which
most of them were smeared, as a protection
against the drying influence of the sun. Unless
therefore, one went before sunrise to the bath,
or after sunset, and thus ran the risk of meeting
with a lion, it was difficult to approach the water
at other times with any appearance of decency.

The next object of my attention at the Bath,
was to procure as many pack oxen as I could.
In England it is quite a mistake to suppose that
pack oxen can be readily procured in Africa,

and that a traveller has nothing to do but to land at particular points on the coast, purchase pack oxen, and then placing himself under the protection of one chief after another, thus traverse the interior. This scheme I should say is generally impracticable, though I, like others, at first imagined that there would be few difficulties attending it; but from what I have heard from those who have been about the Mozambique channel, and from what I myself experienced on the west coast, it is exceedingly difficult to induce the natives to part with their pack oxen on reasonable terms. The people are lazy, train as few oxen as they can (to transport their mat huts, cooking things, their children and themselves from place to place,) and though not averse to part with ordinary cattle, yet they hesitate about giving up their pack oxen, especially since some of the wild breeds of Africa are not trained under some weeks.

The first pack ox I bought at the Bath, was after a week's arguing and bargaining, and it was purchased at last for a large printed cotton shawl, a knife, a tinderbox, twelve bullets, and a pound of canister powder. I trained some oxen and purchased others, and got together, with the six I bought, by good fortune, at Lily Fountain, about fourteen head.

To train a young ox for riding, or for a pack, it is thrown on the ground, and a short stick with a fork at one end is thrust through the cartilage of its nose; to the ends of the stick a thong is attached, which forms the bridle; sheep skins are placed on the back and secured with riems or thongs; the ox is then mounted by a good rider, who holds fast by the belly thongs, and allows the ox to plunge with him, or to run off, till it is tired: and thus, after a time, its spirit is broken, though some ugly falls are got, and much scratching is endured among the bushes by the rider, during the process of training.

I now arranged all my stores, and took an account of them; and anxious to know what sport could be furnished in the neighbourhood, I gave charges of powder and ball to eight or ten of Abram's tribe. They were out a whole day and returned empty handed. But I found out that this was a trick of theirs; for there was game near, though these hunters chose rather to keep the ammunition than to expend it for my benefit. However, my own people brought in plenty of honey, which, though sometimes bitter from the strange wild plants from which the bees procured it, yet the bitterness was corrected by mixing it with milk.

Elliot had been out sporting and returned with a full haversack, whilst a number of the dogs of the Bath were running after him, and licking his legs. I could not account for this, till I found that he was loaded with honey, which, unknown to him, had oozed out through the canvass, and ran in streams to his heels.

I walked out, on two or three occasions, some distance to the north of the Bath, towards black and conical hills of clink stone and of two or three hundred feet elevation, and I always found plenty of springboks on the plain.

It is most interesting to watch the movements of the troops of this most beautiful and graceful of antelopes. Of light and airy form and delicate proportions, its general colour is cinnamon on the back, and the breast and belly white, with a broad longitudinal band, on each side, of deep red approaching to black. When the antelope springs, it shows a broad disk of white on the croup, owing to the expansion of a folding skin behind. The small lyrated horns rise perpendicularly from the brows, diverge, and then incline inwards. When the springboks were first disturbed they would trot off a short distance, then turn and gaze, and if danger came nearer, they would commence to run with their heads to the ground, and then make the most strange perpendicular bounds of six or seven or more feet in the air; and when this commenced we knew it was of no use attempting to get near them.

> " In speed
> They sprightly put their faith, and roused by fear
> Give all their swift aerial forms to flight,
> Against the breeze they dart, that way the more
> To leave the lessening murderer's cry behind.

My driver and leader were out one day with their guns, and seeing a young springbok asleep under a bush, Wilhelm (who had hardly ever fired before) made three attempts to shoot it, and thinking he had hit it the last time, he ran up to it and caught it, when it merely awoke from its dose. It was then brought to the station, was fed on milk, soon became very tame, and, under the name of Jack, was a great favourite with my people. Robert in particular it followed like a dog, dancing after him and bounding with little bells about its neck.

I was anxious to ascertain the extent of knowledge among the tribe with which I now dwelt, to learn what they knew of themselves, and of men and things in general; but I must say, that they positively knew nothing beyond tracing game and breaking-in pack-oxen. They did not know one year from another; they only knew that at certain times the trees and flowers bloom, and that then rain was expected. As to their own age, they knew no more what it was than idiots. Some even had no names. Of numbers, of course, they were nearly quite ignorant; few could count above five; and he was a clever fellow who could tell his ten fingers. Above all they had not the least idea of God or of a future state. They were literally like the beasts which perish.

Strange to say, these Namaquas have no word

for thanks, and they never acknowledged a favour. They thought it added to their consequence to become ill and to take medicine. They got chills in their "windmill" huts, through some of which the air had free course; or they got ill by drinking water all day: still they would be up, and off hunting next morning. We found them tolerably honest, though Mr. Cooke, when he first arrived among them, had one of his sheep killed and eaten in the field.

These Namaquas thought that they came from the east. In the country there is occasionally found (besides the common graves covered with a heap of stones) large heaps of stones, on which had been thrown a few bushes; and if the Namaquas are asked what these are, they say that Heijé Eibib, their great Father is below the heap: they do not know what he is like, or what he does; they only imagine he also came from the east, and had plenty of sheep and goats; and when they add a stone or branch to the heap, they mutter "Give us plenty of cattle."

The Namaquas believe in witchcraft; but this superstition had not the influence I found it had

among the Caffers of the east coast: still the
Namaquas have charms to produce certain effects. Thus some of the people under Mr.
Schmelen had been with the Bondelzwart tribe,
and had got various articles for practising witchcraft. Mr. Schmelen heard of this, and wishing
to get possession of these things for the purpose
of destroying them, he sent for the person who
had them, who came with several others, and
brought a small skin sack with him. Mr. Schmelen, with pen, ink, and paper before him, took a
bit of wood out of the sack, and gravely asked
what that was good for.

" If one chews that," was the answer, " and
then blows towards his enemy, he will prevail
over him."

" Very well," said Mr. Schmelen, and pretended to make an entry on the paper. The
missionary next took out a bone: " What is this
good for?"

" If a person carries this about him, no lion
can hurt him."

This is entered as before. Then a greasy rag
is produced.

" This if put up at a door would bring plenty of wives."

And so on; and when Mr. Schmelen had gone through the list of articles, he asked, " is this all you have?" " Yes." On which he took them up, and threw the whole on the fire, when a blue flame arose from them, and they were entirely consumed. " You see that flame," said Mr. Schmelen, " that is the sign of hell, of the punishment which awaits those who place trust in witchcraft, and not in God." The owner of the charms was sore distressed at losing them, but he thought that some terrible judgment would fall on Mr. Schmelen for having used them as he had done, and thus he comforted himself.

The sun, by some of the people of this benighted land, is considered to be a mass of fat, which descends nightly to the sea, where it is laid hold of by the chief of a white man's ship, who cuts a portion of tallow off it, and giving it a kick, it bounds away, sinks under the wave, goes round below, and then comes up again in the east next morning, its fat having again grown.

There is a strange story about the moon, which is a little better than their usual ignorant notions. The moon, they say, wished to send a message to men, and the hare said that he would take it. "Run, then," said the moon, " and tell men that as I die, and am renewed, so shall they also be renewed." But the hare deceived men, and said, " As I die and perish, so shall you also." Old Namaquas will not therefore touch hare's flesh: but the young men may partake of it; that is, before the ceremony of making them men is performed, which merely consists in slaughtering and eating an ox or a couple of sheep.

I never saw or heard of a people with fewer ceremonies or observances. They take wives to themselves merely by giving presents to the parents; sometimes two chiefs will have four wives between them: this is, I think, new. When a young woman attains the age of puberty, she is led round the kraal, to touch various things for good luck; thus, she touches the milk bambus in the houses—the rams in the fold. When a person is sick, the doctor comes and orders a good sheep to be killed, as he can do nothing without first eating plenty of fat; he reserves a little of the fat to smear the patient with, or he scarifies the flesh over the seat of the disease. When death happens, a hole is dug with a gemsbok's horn or a stick; the body is thrust into it, in a sitting posture; stones are piled over it, and the horn or stick is left upright on the heap.

This, in one paragraph, is a summary of the Namaqua customs. Of course I was not satisfied with this; but I got old men together, gave them tobacco, and cross-questioned them as follows:—

What laws have the Namaquas?

They have none, they only listen to their chiefs.

In the old times used they to sow any grain, or had they gardens?

No; they did nothing of the sort—not before the missionaries showed them how to sow and plant.

What could the Namaquas make before the missionaries came to the Great River?

They could soften skins for their karosses, sew them together with sinews; make bows, arrows, lances, and small axes; bambus for milk; and could weave rush mats.

What is the principal occupation of the men?

Hunting.

How are the women employed?

They put up the mat huts, soften skins, weave mats, and prepare the victuals.   If they decline work, they get the strap.

How is a chief chosen?

The eldest son of the last chief is selected.

How do the chiefs choose their wives?

Any how; from their own place, or from that of their neighbours.

How much is paid for a wife commonly?

From ten oxen to ten sheep, to the father of the girl; and if she is an orphan, her brother gets the amount of her price.

Is circumcision practised in Namaqua land?

No, not at all.

Do the people know any thing of the stars?

Nothing.

Who is the greatest hunter here?

When a lion has to be killed, the chief must go out and endeavour to destroy it.

Where did the Namaquas first get iron?

We think we got it from the east before we saw white men.

Do the Namaquas believe in lucky and un-lucky days, omens, &c.?

They dont know anything of these things.

Are there rainmakers in the land?

None.

Do the people assemble in council or petso, as the Bechuanas do?

No, the chief merely talks with his head men on any difficult case.

Has the captain any particular piece of the ox reserved for him?

No particular part.

Of what use have the missionaries been to the people about the Great River?

Before the missionaries came, the people knew nothing at all; they lived without any thought; they had no worship; all they cared for were their wives, children, cattle, and sheep.

What do the old Namaquas think becomes of people when they die?

They know nothing of these things; all they

see is, that people die and are buried, but what becomes of them they know not; and before the missionaries came to the Great River, the people had never heard of another world.

What had the Namaquas the most pleasure in, their women, tobacco, cattle, beads, or what?

(After some hesitation.) They thought more of their sheep than of any thing else; of tobacco they knew nothing some years ago; it was brought first from the south side of the Great River; and now having tasted it, they prefer it to all things in the world.

What is the worst thing which could happen to a Namaqua?

The death of the sheep.

How did they use their sheep? did they milk them—did they eat them?

They milked them, and sometimes killed one or two when they wanted a kaross. They never killed them if they had anything else to eat.

What do the Namaquas think of white men in general?

The first time we saw white men we thought they were "angry things" that would hurt us;

but after we heard the Word of God we thought that the white men were better than ourselves, or that they were above us.

Is there a great difference in the country with regard to wild beasts, within the recollection of the old men?

Yes: there were more lions formerly in this district. We killed a number with our assegaes.—

One day, a man came to me with a great hole in his arm from a lion bite. I asked him how he had got it; and he said, that some time ago he had been hunting bucks to the eastward, and at a time when his gun was unloaded, he saw, all at once, a troop of six lions close to him. He was on horseback, and did all he could to get his horse out of the way; but it would not move from terror. He accordingly jumped off; when the horse leapt over one bush and he went over another, one of the lions following him. Knowing that it is of no use to attempt to run away from a lion, he turned round and faced him, the animal standing within five yards of him, glaring at him and growling like an angry dog. He then thought of loading his gun; but the moment the lion heard the creak of the stopper of the powder-horn, it flew at him and got his left arm in its mouth, and held him fast. He then felt for his knife; but, on opening it, finding that the point was blunt, he knew it was of no use trying to stab with it. He then took his sambok, or whip of rhinoceros' hide, which hung from his wrist, and hit the lion sharply over the head with it. Another lion now appeared. " Now," thought he, " I am a dead man." But the first lion retired, and the second gave him a blow on the shoulder which threw him on the ground. This one also left him. He then gathered himself up the best way he could, with a crushed arm — out of which splinters of bone came for some time; till at last it healed up, and left only the hole which I saw.

On Sundays, at the Bath, I hoisted the union jack on the waggon. After breakfast, Mr. Jackson preached in Dutch to a crowded Namaqua congregation, and his sermon was interpreted, sentence by sentence, into the Namaqua language, by a native schoolmaster. The people were fond of singing, though their voices were rather shrill. Mr. Jackson, assisted by Mrs. Jackson and the schoolmaster, taught a school of children, on week days, from the Dutch bible. Mr. Jackson was a young and a zealous missionary. His situation in the wilderness—two hundred miles from Lily Fountain, the next Wesleyan station, and amidst a tribe bearing a bad reputation for treachery, and to which people he paid dear for what articles of food he wanted —was not to be envied. Besides this, the heat reflected from the sand and the grey granite rocks, is excessive at the Bath in December, January, February, and March. In the beginning of December the thermometer was generally 80° at mid day.

I practised my men at the target as before, and they improved rapidly. The natives, also, when they had a tolerable supply of powder, were constantly trying their guns; and, strange to say, they made a gun shoot straight which might not have done so when they first got it. This was effected by placing a second sight near the breech, which was raised or depressed, shifted

from one side to the other, and then fixed when
the ball was found to go fair to the mark. One
man had bought from a whaler a bag full of
gunpowder for some oxen, but when he came to
try it he found it to be little better than char-
coal!

I had reserved a few bottles of brandy for my
men, and on Saturday night punch was made,
and I encouraged them in singing and telling
stories, and thus promoted good fellowship among
them. Once they danced; but as this is against
Wesleyan rules of decorum, I did not allow it to
be repeated. On Sunday evenings Mr. Jackson
kindly gave them a lecture.

Besides the occupation of shooting birds for
the purpose of preserving their skins, I made a
few excursions in the neighbourhood to see the
country, and to prevent my people wearying, till
the thunder rains should enable us to move onwards. Our first excursion was to Elliot's Hill,
a few miles north the Bath, which was conical,
two or three hundred feet high, and composed
of black and shining clink stone. From its
summit a common African prospect was obtained.

A wide sandy plain, on which the grass was at this season as white as hay, and where a solitary ostrich pecked the tops of the shrubs. The course of a river, marked by tamarisk and acacia trees along its banks, and in the distance blue mountains.

We went after springboks and ostriches, and in the afternoon, on returning to the waggon, under the hill, we were surrounded with bees in great numbers, anxious to drink from our small water casks and canteens. It was distressing to see how thirsty the little things were.

On the 18th December, the thermometer was 104°—wind northerly.

Our next excursion was to Africaner's kraal, several days' journey E.S.E. of us. Old Africaner was a celebrated robber chief, and used, with his band of desperadoes and guns, to be the terror of the district of the Orange river. He had once paid the Bath a visit, having quarrelled with the Bondelzwarts, and had burnt and destroyed the place, carrying off also some cattle. Old Africaner was now dead, but one of the sons lived at his father's kraal.

I set out with Mr. Jackson and four or five men, in the waggon, containing only our guns, karosses for sleeping in, and a small store of provisions. On inspanning we had a battle with a young ox, which would not submit its neck to the yoke; it was held by the horns and one leg with thongs, yet it kicked, bellowed, lay down, and at last broke away, and chased me round a tree; but at last we mastered it.

We travelled E. and then S.E., over the plain, during a clear night. The dogs ever and anon ran out barking at wild animals which we could not see; at last they got hold of something, and in running up to assist, we found them engaged with two black and grey striped polecats, with long bushy tails, which were exceedingly difficult to kill, and which emitted a horrid smell.

Again the dogs went in pursuit, when two large objects approached us, one dark, the other white. Now there had been a good deal of discourse about white lions, which are occasionally seen in the land, and which we were exceedingly anxious to get, (part of the paw of one I saw). "Here's the white lion!" cries

·Elliott. " Hand me my rifle, I'll shoot him first and charge him afterwards." Elliott was just going to fire, with his rifle sword fixed, when I prevented him till we ascertained positively what the two objects were, when a clear neigh indicated my grey horse, old Night, who was now nearly shot for a curiosity.

We outspanned after eight hours' travel, and the sun at rising bore 150° over the plain on the 20th of December, thermometer 65°. We again inspanned, and after six hours' rapid travelling, *i. e.*, at the rate of four miles an hour with our waggon, we reached the fountain of 'Kururu (the noisy), over which hung a willow below some crags. It was a very lionish looking place.

Next day we travelled E.S.E. six and a half hours, at the rate of three miles an hour, between hills and over a broken country, and outspanned on a height, where we made large fires for one of the party who had strayed; and travelling again in the same direction on the 22d for five and a half hours, we saw the Great River below us, bounded by its billow-like mountains and hills, a scene of the wildest grandeur; and we outspanned at a fountain called Naros (lizard), where were two or three huts. We saw here a cow with a walking-stick; that is, being wild, a long stick was thrust through the cartilage of the nose, with a fork at one end of it to prevent its slipping out: the stick nearly touched the ground as she walked.

Three and a half hours more E.N.E., up a water-course and between hills, brought us to some bushes convenient for our resting-place, and on the 23d, after four hours travel, we descended to a valley with plenty of trees and brushwood, and in the distance were picturesque hills. Here, beside a good garden full of tobacco, maize, and melons, and watered by two fountains, were the four huts of the people of Africaner the younger, about one hundred miles east of the Warm Bath.

The chief was absent; but his people supplied us with a little milk for tobacco; and Mr. Jackson exhorted them and reasoned with them on the sinfulness of making forays on the Damaras to plunder them of cattle, of beer-drinking, of having more wives than one; and though Mr. Jackson spoke very plainly and fearlessly, the people heard all he said very patiently, which I did not expect.

We got here plenty of birds, as white Egyptian vultures, small green and crimson parrots, colleys with blueish plumage, crests, and long tails, black sprews or thrushes, and doves of various colours. The thermometer being at blood heat (98°) at mid day, the temperature was rather exhausting; yet we had altogether a pleasant outspan under the trees by the side of a pool of water.

On the 24th we turned by another route towards the Bath, and came first to a fountain among hills where the reeds were twelve feet high, and then slept in the dry bed of a river on Christmas eve, when we made large fires to keep off the lions.

On Christmas-day we were forced to travel to a brack water; where Mr. Jackson gave us a discourse on sin; and again inspanning, we traversed a great flat, where I killed a yellow cobra capello snake, five feet long, and a horned cerastes two feet long. I now saw, for the first time, one of those extraordinary nests resembling a small hay-stack, built by republican birds on an old tree. The little brown and speckle-headed tenants issued from it in flights, or were carrying straw in their beaks, or seeds for their young, whilst some sat in a row on one of the branches. The pretty little parrots, before noticed, sometimes take possession, perhaps always, of some of the cells of the republicans to breed in.

Near this great nest I tapped a decayed and fallen tree, and out of its trunk I got a good supply of honeycomb. This assisted our Christmas fare, which, eaten on an open plain at night, by the light of a small fire, consisted of some pieces of mutton roasted on the ashes, two or three onions, a few biscuits and coffee; and though we had neither "goose nor gravy," we felt contented and happy.

> " Hail to the eve! when a welcome we give
>   To the friends we hold most dear;
> When we laugh at the jest of each happy guest,
>   Who partakes our homely cheer!
> On this joyous night all hearts should be bright,
>   On earth no soul be mourning;
> While innocent mirth encircles the hearth,
>   When the Christmas log,
>   When the Christmas log!
>   When the Christmas log is burning!

We passed through plenty of grass, but it was so dry that the oxen could not eat it. We were ten hours inspanned at a stretch, then outspanned for an hour in the bed of a river, and went on three hours more to the Bath, which we reached with the poor cattle dying of thirst. They had not tasted water for fifty hours, and trotted off to the water the moment they were released from the yoke, drank their fill at the hot spring, and then lay down to sleep.

CHAPTER VIII.

New Year's Day at the Warm Bath — The Egyptian Vulture — Choubib the Interpreter — His Story and that of Old Balli — More of Namaqua Peculiarities — The Wild Animals found in their Country — The Disposition of the People — Their Huts — Abram's Country and People — His Neighbours — An Evening Sketch — Notions of Chastity — The Chief is displeased with the Missionary — Excursion to Twanos — Rough Riders — Prepare to leave the Bath — A Panic — Prosecute our Journey Northwards — Lose the Cattle — Good Fortune — Dubbee Knabies — Knabies — Kurekhas — A Lion Story — Cowardice of the Escort — Kanus — Leave the Waggon for the Robber Henrick's Place — Aribanies — The Karas Mountains — Kama-Kams — Send for the Robber — Singular Conference with Him and his Mother — Expected Skirmish — Wild Horses — Return to the Waggon.

ON the 1st of January, 1837, we were still anxiously looking out for rain at the Bath, as the grass, from the heat of the sun, was beginning to blacken about the roots, and there was a decrease in the water of the hot spring. Thermometer at mid day 93°. In the afternoon the sky was overcast in the direction of our further journey, thunder growled towards the north, clouds of dust swept across the plain, and in the evening we had a shower, accompanied with a beautiful double rainbow.

Still we heard no good news of the state of the country on our proposed route, and we were compelled to tarry a little longer at the Bath.

My people generally kept their health, though the heat was often intense. Thermometer on two or three occasions as high as 110°. Taylor for some days suffered severely from inflammation in the kidneys; and I thought, one day, I should have lost him, but with calomel and opium a cure was effected.

We were very anxious to secure a pair of Egyptian vultures, which young and old, brown and white, in considerable numbers, constantly careered in circles above the huts of the kraal, or sat on the grey rocks near. My people for many days tried to shoot one, but without success. Buck shot and even pistol bullets were used in vain; for though the birds were sometimes knocked over, yet they always rose again, and escaped. At last, by crawling on my hands and knees, towards a dead horse, I got within forty yards of a gorged vulture in full plumage, and penetrated his tough hide with No. 1. This shot my people thought would break the spell of our detention.

I now engaged an interpreter for the journey in rather a singular way. An old man, Choubib, who could speak Dutch, came from a distance to complain of Henrick, a captain under Abram, by whom he had been robbed. Choubib's story was shortly this:—his brother-in-law, a Bastaard of the name of Engelbreght, was out hunting a year before this with Henrick's father, and whilst they were resting themselves and smoking together at a bush, a troop of zebras galloped past. Engelbreght hastily snatched up his gun; it had got entangled with a branch, went off accidentally, and old Henrick was shot through the body, and shortly after expired, merely requesting that his friends would come and bury him. Engelbreght, knowing the vindictiveness of old Henrick's people, fled for protection to Choubib.

Young Henrick sent to demand that Engel-

breght should be given up, that with his life he
might pay the price of blood for blood; but
Choubib would not surrender him; and said he
should take Engelbreght to the Bath, to be
judged there by the chief and the missionary. He
did so, and old Henrick's death was found to
have happened by accident. Choubib and his
brother-in-law returned home; and one day
whilst he and Engelbreght were out hunting,
Henrick sent a commando against Choubib's
kraal; the women were plundered of their beads
and skins, and otherwise ill used; the herd was
killed in the field with stones, and thirty head of
cattle, forty sheep and goats, two guns, and some
horses, were carried off. In such a lawless state
is Great Namaqua land.

Choubib said that he and his people were
starving. Abram, with his usual dogged indo-
lence, did not listen to the tale of injustice. I
supported Choubib with food for several days,
and then went to Abram with him, and persuaded
the chief to send off three men to demand Choubib's property; when the old man said he would
guide me to Walvisch Bay, if I chose. I was

very glad to have his services, for which he asked
a gun and some ammunition.

I shall give another case of the sort of justice
which prevails in the land. Old Balli, whom we
had before seen at Henkrees, now came to the
Bath with two of his sons with this complaint.
Seven years before, he had given in charge of
Marcus, one of Abram's men, one hundred and
twenty-six sheep, that they might be fed and
cared for—Marcus receiving payment from their
produce for his trouble. But Marcus going on
a hunt towards the Damara country, lost all the
sheep but forty-six. He said he should give four
oxen for the remainder, to which Balli agreed;
but Marcus altering his mind, forced a gun
upon Balli for the missing sheep. Then Abram,
being at war with Africaner, said to Balli,
" You don't want a gun now; I do, lend me the
gun." Balli did so.

Subsequently, Balli's sons were at the Bath,
to endeavour to get the gun. Abram abused
them and threatened them. They fled by night,
in fear of their lives. Abram then sent for Balli
himself; but he being afraid, declined coming

among the treacherous Bondelzwarts. Abram's messenger then robbed him of two horses, which he found in the field, to make him come. Poor Balli thinking that with my party at the Bath, he might now be safe, came to me. I interfered in his behalf, and Abram sent messengers to get the horses, and promised to see the old man righted.

Not to tire the reader with too much of one subject at a time, I did not say at once, all I intended to give regarding the peculiarities of the Namaquas. I now give some more of their usages, and a sketch of their country, believing that those who feel an interest in the various members of the great human family, will be better pleased with these slight details regarding a primitive race, than with minute descriptions of lower animals and plants.

The Great Namaquas may be said to extend along the 'Oup or Great Fish river, on both sides of it, and to occupy at different seasons its banks and those of the numerous streams which fall into it. Some districts of the country of the Great Namaquas, especially those east of the

Great Fish river, in its upper part, produce plenty of grass for large herds of cattle; while beyond the sources of the Fish river, the country of the Kamaka or Cattle Damaras, or Damaras of the Plains, is very fertile, the plains being grassy and full of cattle, whereas the Damaras of the Hills inhabit a region which is not generally adapted for grazing. Many parts of the Great Namaqua country are also very barren and mountainous.

All the large wild animals are to be found in the Namaqua country; but elephants are now several days' journey east of the Fish river. Lions are every where found; most of which are of the usual light brown colour, whilst others are entirely black, with long hair; a third sort is white; a fourth has striped legs, like those of a tiger; and a fifth has a white neck. I saw the common lion and part of a white one, the others I heard of from the natives, and I feel confident that they exist. Two-horned rhinoceroses, both black and white, are now found in the upper parts of the Fish river; zebras are every where in the land; beautiful spotted panthers; plenty of

giraffes or cameleopards, buffaloes, koodoos, gem-
boks, elands, hartibeests, klip-springers, springboks, and others of the deer tribe; hyenas, wild
boars, jackals, polecats, rats and mice, are in great
abundance.

The larger birds are ostriches, eagles, vultures,
bustards, cranes, pheasants, and guinea fowl.
There is a great variety of small birds, particularly along the constant waters of the Fish river.
Snakes and serpents are in plenty; but fortunately
there are very few mosquitoes. The people are
not much tormented with the plague of insects; except with those which multiply from habits of neglect, and with a troublesome and small red bodied
tormentor with eight legs, called a bush louse.

The Great Namaquas are taller than the Little
Namaquas, but have the same general resem-
blances, their colour being yellowish brown, hair
crisp and curled, noses and eyes small, face
triangular, lips protruding. Both sexes are fond
of greasing the skin, and the women also bookoo
themselves, that is, they rub the ground root of
the bookoo plant, which has an agreeable smell,
over their persons, and sometimes draw odd

looking streaks of soot and grease on their faces.

The Great Namaquas use the very same clicking dialect as the Little Namaquas do. Almost every word has an initial click, or has one in the middle of it, and some words have two clicks. The clicks are of three kinds: one is performed by striking the tongue against the palate and front teeth; another by striking the tongue against the centre of the roof of the mouth; and a third by striking the tongue far back in the mouth. The word *'un'uma* (bulb) is an example of a word with two clicks (') in it. I need hardly add that the language is one of great difficulty for a stranger to acquire and pronounce; the clicks resembling one another so closely, and each conveying a different signification.

The Great Namaquas are not a bloodthirsty people, though their fondness for cattle induces them occasionally to attack the Damaras of the Plains. Strangers who visit the Namaquas, are generally treated with kindness, and he is held in great contempt who eats, drinks, or smokes alone. However, the murder of Mr. Trelfall shows that a traveller in Great Namaqua land ought not to trust too much to the forbearance of the people, if they think they can easily overpower him, or may not be afterwards called to account for his death.

The huts are universally composed of bent boughs, covered with neatly woven mats, and are perfect hemispheres. These huts are easily removed from one place to another: the mats are rolled up and tied along with the boughs on the backs of oxen, the earthen cooking pots and milk bambus hang from forked sticks on each side, and the children, two or three, one behind the other, sit astride of the ox, and hold on by the upright sticks; the mother drives the ox, which is laden with her offspring, her house, and utensils.

The people at the Bath amounted to between five or six hundred souls; but these were not all the adherents of Abram; the others lay at different places, some distance from the Bath: perhaps his people may amount to two or three thousand souls.

Abram's country may be said to extend one hundred and eighty miles north of the Orange

river, and it is about one hundred miles broad.
The Chief Kuisip is to the west of him; Amral
to the north-east; the Africaners to the east; to
the north-west are the Buys of Bethany; west is
Kurusumop, and Paul Lynx is at the mouth of
the Orange river.

Abram's people had plenty of cattle, sheep,
and goats among them. Clouds of dust rose on
all sides in the evening: the flocks and herds
then appeared; and the lambs and kids, which had
been confined at the kraal all day, ran out noisely
to meet their mothers. The women secured the
cows, and bringing the calves to them, allowed
them to suck a little, and then appropriated the
rest of the milk. The cows, whose calves had
died, were blown in an extraordinary manner, to
make them let down their milk, as described by
Kolben in his strange work on the Cape. After
sunset the men assembled, generally at my waggon, to talk about hunting, and to smoke. I
got up wrestling matches among the young men,
and they showed great agility in this exercise,
though my boy Antonio contended with them
successfully.

During the night, I am sorry to say, there was a good deal of visiting for no good purposes; and figures were often seen moving from one hut to the other. Chastity is of small account among the Namaquas: the chiefs even when they go to the sea, lend their wives to the white men for cotton handkerchiefs, or brandy; and if a husband has been out hunting, and on his return finds his place occupied, he sits down at the door of his hut, and the paramour handing him out a bit of tobacco, the injured man contentedly smokes it till the other chooses to retire. This surely is the acmé of complaisance.

I have already noticed the dress, arms, and ornaments of the Little Namaquas, which resemble those of the Great Namaquas.

Abram was for some days in a very bad humour with Mr. Jackson, because the missionary had been lecturing him about certain neglects of duty, and had " put his finger in the chief's eye," as Abram expressed it. At last he came round. The messengers he had sent to Henrick on Choubib's account now returned, and reported that if Choubib wanted his property, he might come and fetch it himself; that Henrick would not send it; and that Abram had not yet comforted the hearts of Henrick and his people, the 'Haboobees, or " leather shoe wearers," for the death of Henrick's father. I now said that I wished Abram to go with me to Henrick, when we left the Bath, to endeavour to recover Choubib my interpreter's property, and Abram agreed to go with me. I let Choubib go back to his people for a few days.

The riding oxen for my people having been now trained, I was desirous of trying them, and accordingly set off on another excursion, eighteen miles to the south-east, to a place among hills, called Twanos, or " run over in a morning." I took the chief with me to put him in a good humour, and three or four of my people. At Twanos there were only two huts, in a wild glen, and before them was a garden, with one pumpkin in it and six stalks of tobacco !

We fired at a mark, took our coffee and karbonatjé (or pieces of meat roasted on twigs) in the evening, and slept under a rock on the sand, the Namaqua feather bed. Sand is preferable to grass for a bed, though the gipsies sing,

" Our banquets may be coarse and rude
    Our board be spread in open air,
But we can relish humble food
    While others loathe a richer fare.

Let wealth in perfumed chamber lie,
    Let damask guard his fevered rest,
Our curtain is the starry sky,
    Our couch the green earth's dewy breast."

Next morning the chief drank honey beer, and became very facetious, talking incessantly all the way home of his skill in shooting lions !

This Namaqua notion was curious. The brother of the chief had put out his arm when out hunting. I lubricated it for a week with grease, and then strapping the patient on a cot, I with two of my men drew the arm with all our might to reset it. We much improved it; when the Namaqua coolly asked a present for having allowed himself to be operated on.

Some of the people now killed four horses on an ostrich hunt. I never saw horsemen ride so unmercifully as the Namaquas do. I lost my own horses for several days: England strayed as far as the Great River, and would probably have revisited his accustomed pastures under the

Kamiesberg, had the Great River not frightened
him with a flood.

Seeing for two or three days in succession
heavy masses of clouds to the north, though we
had no rain at the Bath, I determined to quit it
at all hazards; and accordingly got a light waggon
(exchanged for the nonce with Mr. Jackson)
greased and in order, and packs prepared for six
bullocks; and I told Abram that I should expect
him to accompany me with a party to the next
chief, and that we must leave on the 18th of the
month.

The 18th arrived. With great difficulty I had
collected some more sheep. From Abram I had
got no assistance, though one would have thought,
that for his annual present of powder and lead
from the colonial government, he would have
either offered me a present, according to Namaqua custom, or been active in getting sheep
and other things I wanted to purchase; but no,
he was brutally sulky and bad hearted, as usual.
The waggon was inspanned, oxen packed, and
every thing in order to start, when at sundown
I observed a grand consultation among the Bon-

delzwarts. The chief and the twelve men of the escort, with their guns in their hands, and ball pouches strapped round them, stood, with others of the tribe, in a circle, and were speaking earnestly. Mr. Jackson went to them, and returned with this news: that the people were afraid to go with me; that a young man had arrived from the north, and had told them that an old Bastaard, of the name of Dirk, with whom they had had a quarrel some years before, had removed, with his cattle and sheep, out of our way, hearing that we were coming, and that therefore Abram was afraid of mischief.

I went to Abram and said, " If your enemy goes out of the way, that is the very reason we ought to proceed."—Abram : " But if everybody goes out of our way, who will show the water places ?"—I answered, that my interpreter, Choubib, who would meet us, knew the country well. —" Wait till to-morrow," said the *gallant* chief, " and let us hear more of this news."—I replied that I would on no account wait; that if we were to believe every Namaqua lie, it was of little use my leaving the Cape; and that now, if the chief and his people would not go with me, according to agreement, I would go alone, if I could get a guide.—After a great deal of difficulty, and the offer of a present of the all-powerful weed, at nine o'clock I got a guide, but he said he had no shoes. I said, in joke, "take mine;" and he proceeded to do so: however, he got the loan of another pair, and we started. My people were highly indignant at the cowardice now exhibited by the Bondelzwarts, and one of them gave an Irish receipt to make them behave themselves, which was "to clap a musket to their breast and threaten to blow their brains out."

I parted from Mr. and Mrs. Jackson (the latter of whom had a baby at the breast) with many thanks for their civility and kindness to me and to my people; and shouldering my old companion in the East, a double-barrelled Manton, I stalked after my party. It was moonlight; but having fewer drivers than I expected for my loose oxen, one disappeared after another, among the trees and bushes of the 'Hoom river, up which our route lay; and at two in the morning, half of the sixteen loose oxen were missing, and most of the packed ones. The waggon was halted and out-

spanned, the people went out in search of the
cattle, and two men went back to the Bath on
my horses. Till daylight I was forced to endure
this bad beginning with as much patience and
philosophy as I could muster; and after rambling
about for some time in a vain search, I lay down
under a bush, covered with the leopard's skin of
my saddle; but it was impossible to sleep for
thinking of the strayed cattle and the disgraceful
conduct of the escort:—altogether my situation
was most uncomfortable.

Daylight came, and with it the discovery of
some of the pack oxen from a neighbouring hill.
I inspanned, and went towards the next water;
when, by-and-bye, Abram and his fighting men
joined me, on their riding oxen; said no more
about Dirk, and I did not ask about him; and
we arrived at the water, at Dubbee Knabies, the
place of Dubbee or tamarisk trees, six hours
from the Bath. Here I fortunately recovered
all my runaways. However my good luck did
not end here: the atmosphere became close and
stifling; thermometer at 95°; clouds gathered
overhead; we had a very heavy shower of rain,
which enlivened in a wonderful manner both men

and cattle. But we observed that the rain cloud passed by the unfortunate Bath, as usual.

The water place was in the bed of the 'Hoom, and was cut no less than ten feet deep. To it the impatient cattle and sheep were allowed to descend in squads, to prevent the body of them trampling in the sand: so painfully is the indispensible element obtained in this land of the sun.

At Dubbee Knabies we saw two or three huts and plenty of cattle and sheep of Abram's and his people. Some of the cattle had the Damara mark upon them, viz., a deep cut in the dewlap; of course I winked hard at this, like an old traveller. Next morning I awoke my people with the bugle before day, and we journeyed onwards. My party had altogether a wild look, with hats covered with ostrich feathers, and jackal tails (the Namaqua handkerchief) stuck in the muzzles of the guns. The packs were troublesome at first; but my people soon got into the way of managing them, and we progressed to the north cheerfully.

Six hours more brought us to another cattle place, Knabies (the place,) from whence we saw a range of mountains to the east of us. We slept under some noble trees; the night was cool; there was plenty of water (a foot from the surface,) and of grass for the cattle. Next morning we lost the cattle from the carelessness of the herds, and we did not find them for many hours.

Two hours brought us to Kurekhas, where there was a reedy fountain, with the foot marks of zebras beside it. Kurekhas was a favourite resort of lions.

The Namaquas sometimes display wonderful intrepidity and desperate courage in the attack of lions. Thus I saw a man of the name of Lynx, who had one hand crushed by a lion in the following encounter. The lion had killed one of his cattle: now, it is the custom in the land to let lions alone if they do not destroy persons or property; but if they do, it is always understood that a hunt must take place: so Lynx, with three of his cousins, now went forth to attack " the governor." They tracked him to a bush, and were preparing for the encounter with their guns, when the lion rushed out unexpectedly; on which the three cousins fled, leaving Lynx, who was immediately overthrown by the lion. The monster seizing him by the left arm, was dragging him off to destroy him at his leisure, when the runaways seeing the predicament in which Lynx was in, immediately turned, but being afraid to fire for fear of hurting Lynx, one jumped on the lion's back and pulled stoutly at his ears, to make him let go his hold, whilst another hung on by the tail, to stop him; the third, watching his opportunity, sent a ball through the animal's forehead; the lion then quitted the arm, and in his death agony he crushed Lynx's hand with his teeth. The dying bite of a beast of prey is always the worst. The above would make a fine group for the inimitable Landseer.

Continuing our course up the 'Hoom, we traversed a great plain close to it, and saw the mountains of Karas or Sharp, bearing N.N.E. of us, thirty or forty miles distant, and two or three thousand feet high. In the evening, though the moon was up, the wind cool, and the time most favourable for pushing forward, the Bondelzwarts said, " This is an angry place for lions, we must outspan;" and so these cowards, " making

lions of mice, and Boschmans out of baboons,"
forced us to halt for the night, and to make
kraals or fences for lions, the traces or voices of
which we had not even seen or heard.*

Next day we reached Kanus, the place of the
Kan Bush; also on the old 'Hoom, or _Hum;_ for
though it was called a river, there was no water
in it apparently, until we dug deep into its bed.

At Kanus I left Taylor and ten men in charge
of the waggon and baggage, and with Robert
Elliot, Abram, and Choubib, who had just rejoined us, and nine of the escort, I set out for
the robber Henrick's place among the recesses
of the Karas mountains, to endeavour to recover
Choubib's cattle.

This was undoubtedly rather a hazardous undertaking, knowing, as I did, the bad character
I had to deal with; still, for the sake of securing
the services of Choubib to the sea, and having
pledged myself to assist him on this occasion, it
was necessary I should go through with the business, and run some risk.

* For the estimated distances see the Itinerary at the end
of the work.

The banks of the 'Hoom, up the bed of which we first passed, consisted of two walls of perpendicular cliffs. Leaving the 'Hoom, we travelled east, and sixteen miles brought us to Aribanies, another cattle place of Abram's, where we got water, like that from a common sewer, for our coffee: I passed the night in the lee of a bush.

Next morning, three hours over a plain, brought us among the Karas mountains: they are tossed up into various summits (the highest * may be three thousand feet above the plains,) and occupy many miles of the superficies of Great Namaqua land. Their basis is granite, as usual. Among the glens, the euphorbia candelabra predominated. After one and twenty miles ride under a scorching sun, we off-saddled in the bed of Kei Kap (Witch R.), whose course is past Africaner's kraal to the Orange river. We were parched with thirst—could find no water— our small canteens were exhausted, and after resting the cattle, we saddled-up again and rode on. Seeing some indications of water in the

* Now named after the general commanding in chief.

bed of the river, and the riding oxen beginning to smell the sand, we anxiously jumped off, and dug with our hands, and found a small supply of the precious element. The trees also dropped gum, which we greedily chewed.

Through most rugged and stony glens (road there was none,) my poor horses and the oxen slipping over the stones, and our clothes torn with thorn bushes, we reached a small hamlet of huts; here, an old woman, an acquaintance of Choubibs, came out, seized his hands, hung her head to one side, whined and cried, and ran after him. Ascending a hill, we off-saddled on the other side, in a narrow valley, near some more huts, at a place called Kama Kams, about fifty miles from where I had left the waggon, and only two from the place of the robber Henrick.

Abram had a right to call Henrick to him, as Henrick was a captain under the chief: two men were accordingly dispatched to Henrick with this message, that a white captain wished to see him, and to speak to him about sheep and cattle for purchase, and about other matters: he returned for answer, that he should come in the morning.

Henrick appeared at the appointed time—a strutting little fellow, with a long stick in his hand. He was accompanied by a few men: and we sat down under a tree. I asked him if he could sell me any cattle or sheep for my journey, and he said he could spare me only two sheep for cotton handkerchiefs. I then explained why I came into the land:—to see it and to ascertain if a trade could be opened with the people. I then (keeping Choubib out of sight) went over the manner of old Henrick's death, and the seizure of Choubib's cattle and sheep after it; and that the death had been proved at the Bath to have been occasioned by accident. I said that I interfered in this matter because, among other reasons, I wished to see peace in the land, and not war, which there would certainly be if the plundered property was not now given up. I explained that in other countries it is the custom to pay a fine for blood, even if it was shed, as in this case, by accident; that now Henrick would probably get some cattle for the death of his father; but that it was cruel to bring Choubib's people to the verge of starvation by depriving them of all their milk—their chief support; and I added, that we wished to trade with the Namaquas; but that if there was war in the land, our traders could not venture near it. Missionaries also, whom many of the Namaquas desired to have, could not live in the land if the people were fighting with each other;—that if the Namaquas quarrelled among themselves they would be rendered weak, when the Damaras might come down to destroy them;—that as for the English, they were not afraid of Namaquas, Damaras, or of any people in the world;— that we had such a quantity of guns and ammunition, that no people could hurt us; but that in these times, notwithstanding our great power, we never oppressed any one, and instead of our allowing, as in the old times of the Cape Government, the natives of the country to be deprived of their land, the present Governor of the Cape was giving the Hottentots land wherever he could find it vacant in the colony.

To all this Henrick said, " My heart burns for the life of Engelbreght, because he shot my father." I answered that Choubib was to be

praised for not surrendering his brother-in-law—
that he could not have done it—and that if his
property was not given up, he would call on
the great chief Amral, under whom he stood, to
come down and destroy Henrick and his people,
—and that it was impossible they could escape.
To this Henrick answered, " I don't care; I can
but die;" he then, after three hours' stout argument, said " I'll send my mother to you." He
then retired; and in the evening old Henrick's
widow came (a sturdy old hag,) and we soon saw
that, though her son might be brought to terms,
it was this old "limb of Satan" who was at the
bottom of all the mischief.

" Kill and slay," she cried in a fury. " I'll
listen to nothing;—what do you all know about
the matter? I want Engelbreght's wife to be in
the same state as I am—to be a widow as well as
myself. Why should she have a man more than
I have? We find that you have Choubib here
—give him up to us; if we cannot have Engelbreght's life, we can have Choubib's; and blood
we must have." I said we would sooner give up
our own lives than Choubib's; that he was my
interpreter, and under my protection. The old

haridan, cooling a little, then asked me for some tobacco. I gave her a stick of it, and she went off smoking, though not apparently "a calumet of peace."

Matters looked rather awkward, and it was evident that our arms ought to be in fighting order, in case of accident. We accordingly prepared a half moon screen of bushes in an open part of the narrow valley to sleep behind, and defend ourselves, if necessary; and sending a spy to the huts in front for milk, he discovered that a considerable number of muskets had been just sent from Henrick's place, to be in readiness. I told my two white men that we must prepare to sell our lives as dearly as we could; but that I did not doubt, if we could manage to shoot Henrick, or his mother, or both, the first fire, and then rush in with our swords, that the rest would run off, or give in; and that, in the mean time, it was necessary to keep a good look out during the night.

Abram and his people went to sleep on their arms. I took the first watch from ten to twelve, and my men the watches from twelve to two, and two to four; but we had no interruption; and after sunrise Henrick sent a messenger to say that he wanted to speak to me. I said he might come; when (to intimidate us) he appeared with thirty-three strapping fellows, double our number, and it was only on my own two men I could depend, and not on the cowardly Bondelzwarts.

We sat down again to confer. Henrick said he came to listen. I said he must tell us something, and he replied that if we were not so strong, he would take Choubib, kill him on the spot, and give his body to the crows. Then to pick a quarrel he began to question a servant of Choubib's regarding the death of the old captain (Henrick's father). I got impatient, and interfering, said that I could not spare time to go over the story of the death again—that we had discussed all that yesterday—that Henrick should have neither Engelbreght nor Choubib to murder, and that if he did not send Choubib's property to the waggon now, he should hear from us before long. I then ordered my horses, carried off Choubib, and was followed by the Bondelzwarts—Abram, the chief, having been unable to make Henrick listen to him.

" The leather shoe wearers" did not venture
to follow us, or attempt to capture Choubib.
The line of conduct which I now pursued was
eventually attended with good effects ; and I have
given the details of the conference to afford a
better insight into Namaqua feelings and springs
of action than I could have done in any other
way, or by many pages of narrative.

We off-saddled in the middle of the day, and
then continuing our ride among the hills, the
great footmarks of a camel-leopard were seen,
those of a lion, and of ostriches; when, hearing
a rushing noise above us on the right, I looked
up, and saw a troop of wild horses crossing
rapidly the hill side, and causing the loose stones
to rattle beneath their hard feet.   We dashed
forward to intercept them and fired from the
saddle without effect.   The sight of those beautiful tenants of the wilds, in their state of perfect
liberty, was to me very great enjoyment; and as
I was not particularly hungry at the time, I did
not regret that we had missed those " whose
house is made in the wilderness—the barren
land their dwelling, and the range of the mountains their pasture."

The wild horse, zebra, and quagga, nearly resemble each other: the first is striped all over; the head, neck, body, and legs, to the fetlocks, are covered with dark bands; its feet are hard and compact, for its resort is the stony mountains. The zebra again, an inhabitant of the plains, has larger feet, and though its head, neck, and body are striped, its legs are white; while the quagga has only the head and neck striped.

Leaving the mountains, we reached the open country, and had some excellent sport with innumerable springboks, which danced and bounded on each side of us, scouring across the plains, and springing over the bushes as if they had been struck with battledores. We slept again at Aribanies, the men and bullocks throwing themselves headlong into a small pool of water to drink, after a hot day's ride.

Next day, the 28th of January, I reached the waggon at Kanus, and found all safe.

CHAPTER IX.

On Sunday, the 29th of January, after divine
service—which principally consisted in reading
and expounding a chapter of the Bible, which I
think is more likely to be impressive, though
attended with more trouble, than merely reading
prayers—I had Abram and two of his head men
to dine with me; and as they professed a desire
to return home, though they had not gone a

third of the way yet, which they had agreed to
go, I said they might go back to their fire-sides
on this condition—that Abram immediately sent
another message to Henrick, demanding Choubib's property, and if that message did not
succeed, then Abram was to send a commando
or armed party to take Henrick's guns from him,
which would be the best means of humbling
his proud spirit and of bringing him to terms.
Abram faithfully promised to do all this if
I would allow him to go home. Now the truth
is, I was quite tired of the Bondelzwarts:
they were cowardly, and bad in all things
saving their appetite, which was excellent.
I was really glad of an excuse to get rid of
them; for they were of no manner of use (except
to eat my sheep), and I am certain if we had
been attacked by savage men or wild beasts,
they would have fled to a man, all excepting
Abram's chief councillor Hortman, a worthy man,
and, I think, a sincere Christian.

Hortman said, "If I had been Henrick's
chief, I would not have allowed him (an insolent
boy !) to treat me with contempt in the manner

he did Abram, or to laugh at me. I should have taught him with a sambok (or whip). Abram ought immediately to send a commando to him, and there will then be peace in the land."

On the 30th I gave Abram and his head people some small presents of knives, handkerchiefs, tinder-boxes, &c., and exacting a solemn promise from the chief that Choubib's business should not be neglected, I let the Bondelzwarts return from whence they came. No wonder the *amiable* people were afraid to go on with me; for I subsequently heard every where complaints of them, of their plundering their neighbours of cattle, of their violating the wives of strangers who visited the Bath, of their robbing single travellers, of their having attacked the kraals of those with whom they had quarrelled, and mercilessly burning the huts with the women and children in them, and such like stories.

We inspanned and travelled three hours over a plain in a north-north-westerly direction, and were suddenly assailed with such a violent shower of rain that the whole country ran with water, and we were forced to outspan on a small rise. It was a long time before a fire could be kindled, all the wood was so wet. The usual Namaqua method of making a fire was tried in vain, viz., enclosing a piece of burning tinder loosely in a handful of grass, and swinging it round the head till a flame bursts out. Nothing but a lucifer match and plenty of paper enabled us to dry some wood to commence a blaze with.

Two of my men having neglected to change their wet clothes, got seriously ill with cold and fever. Hot water and medicine brought them round again. Many soldiers and sailors require as much looking after as if they were children.

Next day we passed over three of the sources of the Lion River, and on the following day through a poort or pass* of one of the offsets of the Karas mountains, and reaching a fine plain we descended for five hours gradually to another poort with a conical mountain on the left, and a lofty wall of rock on the right. Here on the

* With mountains to the right and left, named after the Military Secretary and Adjutant-General at the Horse Guards.

L 2

banks of the Kaap River, a branch of the Great Fish River, and under lofty trees stood Choubib's kraal of half a dozen huts, at the place called Chubeechees (black ebony).

We now pitched our tent under the trees, as we were obliged to tarry here for a few days to make arrangements for our future progress; but whether halted or marching, we found it advisable always to rise before the sun, remembering these lines of Scott—

> " Time, stern huntsman! who can baulk,
> Staunch as hound, and fleet as hawk?
> Think of this, and rise with day,
> Gentle lords and ladies gay!"

On Wednesday, the 1st of February, the thermometer at sunrise was 65°, and at mid day 90°. A few fly-catchers, black and yellow orioles, parrots, &c., were got here. In the evening Choubib's people, and also those from another kraal a mile down the river, came to the tent to dance. The fiddle, tambourine, and castenets were again in requisition, and the young people enjoyed themselves; and young and old were in great good humour, without any improprieties being committed.

I think a young Namaqua trained as an opera-dancer would excel in agility any of the European *corps de ballet.*

Old Dirk, of whom the Bondelzwarts said they were so much afraid, now visited us with his two sons, and he laughed heartily when I told him that he and his family were sufficient to frighten the whole tribe of Abram. Dirk, who was a civil old Bastaard, said that if I waited two or three days, he should assist me with a span of oxen, and that he wanted to barter some cattle and sheep for some of my goods.

At Chubeechees the people were very poor; for Henrick had robbed them of almost all their milk; there were, however, a few cows and goats left, and of the milk of these we shared. I had also a cow of my own, and a few milch goats. Standing in need of a shepherd, I observed here two or three fine little Damara boys, black as ebony, and about ten years of age, who had been carried off by the Namaquas during northern forays; and one in particular, who had got the name of Saul, came frequently to assist my groom, and asked to go with me. Poor fellow! he had many marks of the sambok on his legs, and he was besides literally starving under an old woman. In the morning he ascended the gum-trees for his breakfast; at mid day, whilst in the field with his mistress's goats, he hunted lizards under the bushes, dexterously knocked them over with stones or a knobbed stick, then skinned and roasted them for his dinner; whilst in the evening, before he folded his charge, he cunningly sat down behind a milch goat, and drained the milk into his mouth: thus he lived from day to day.

I said to the old woman to whom Saul belonged, "You have two boys, and they are starving; you have nothing to give them."

"That is true," she replied.

"Will you part with Saul?" said I. "I want a shepherd, and the boy wants to go with me."

"You will find him too cunning," returned the old dame.

"I want a clever fellow," I said.

"Very well," she replied. "Give me four cotton handkerchiefs, and he is yours."

"Suppose," said I, "you take two handker-chiefs, and two strings of glass beads?"

"Yes; that will do," and so the bargain was closed; and thus a good specimen of Damara flesh and blood was bought for the value of about four shillings!

Saul remembered of his capture that his father, who had plenty of cattle and sheep, was absent at his tobacco garden, when one night there was a noise outside the huts. Saul went to look if the sheep were all right, when a party of Namaqua horsemen rushed forward, killed some of the Damaras, and carried off the cattle. Saul trying to escape, was caught up and carried off by one of the horsemen.

He was now highly delighted to leave the old lady, to be regularly fed, and to escape the sam-bok. I told him to go and bring his skins, on which he informed me that he had none, saving what he stood in, and that was his own sable hide, with the addition of the usual strap of leather round his waist, from which hung a piece of jackal's skin in front.

Constant exposure to the vicissitudes of the

weather without clothes, hardens the skin of the body like that of the face; still it is difficult to sleep at nights without proper covering.   In cold weather the poor creatures in Namaqua land, who may have no karosses, sit cowering over a fire all night, and merely doze, with their heads on their knees.   Think of this, ye pampered menials of our great metropolis, who give warning if asked to eat cold meat, or if made to sleep on aught but feathers!   It would do you great good for the rest of your lives to climb gum-trees for a month for your food, or to sit for a few rainy nights under a bush, endeavouring to keep alive an expiring fire.

Ladies of Peckham! ye whose compassionate hearts used to feel so much for negroes in the West Indies, and in whose eyes white planters are monsters of cruelty! though you may not have believed all I said some few years ago regarding the well-cared for state of slaves under British masters, under your countrymen, yet believe me now, when I say that I did not purchase Saul to sell him again, or to ill use him. I bought him to tend my small flock, and with a view to his eventual emancipation, and education in England. He is now near you at school.*

When Dirk returned I made Robert my head salesman. Shop was accordingly opened under the fly of the waggon, and gaily flowered shawls, red handkerchiefs, variegated beads, shining knives and tinderboxes, saws, hatchets, &c. were temptingly displayed before the assembled people. Bullocks were driven past us for barter, and sheep dragged forward by the horns. Dirk parted with a fine bullock for a carpenter's brace and twelve bits; a slaughter cow for two choppers; another for a shirt, a handkerchief, knife, and tinderbox; sheep for a tinderbox each; and so on, till we had got as many cattle and sheep as I then wanted. Beer was made, on which Choubib and Dirk got very drunk. On the 9th we managed to get under weigh again, and journeying northerly, kept the long line of the Karas Mountains at some distance from us on our left;

* At Woolwich, with the children of the Rifle Brigade—my brother's (Captain J. Alexander Henderson's) regiment.

L 3

and after sixteen miles we outspanned. It was late; no kraal of bushes was put round the sheep; the jackals got among them, scattered them, and next morning, not one was to be seen: we hunted about in various directions on the foot marks for them, got them in two or three clumps at the distance of some miles from the waggon, panic struck, and some bloody, and sustained a loss of three out of the flock.

We again advanced. But in small matters as in great, misfortunes seldom come single, so Jack the springbok, who had now grown large and handsome, for whom the milk of the cow (and she like other Namaqua cows, only gave two or three pints a-day) was entirely allotted, and who had been the constant delight and amusement of my people with his playful gambols, whilst loitering a little behind on the road, was shot by a man of Choubib's, who mistook him for a wild springbok. I believe there were tears shed when poor Jack was carried to the waggon, dying from a large hole in his flank.

We reached the fountain of the Broken Liver,

on a plain among some hills; here we lost all the goats, but recovered them again after many hours pursuit over the hills; and continuing our journey across undulating plains, on the 12th of February we reached the banks of the Kamop or Lion river, where we found the huts of an old Bastaard (with two wives) named Arnoldus.

We were now at the place called Nanebis, to which Abram ought to have brought us, as it was the usual residence of the Chief Kuisip. The chief was not now here, though some of his people lay two miles up the river, and I dispatched a messenger to him at the 'Oup or Great Fish river.

Choubib's brother now joined us with two other men, and brought the bad news that Henrick the robber, had declared he did not care for the Bondelzwarts, and that it was believed he was preparing to attack Choubib's kraal again, to carry off all that remained of his property, and probably murder his wife and children, and then follow and attack us, because I protected Choubib, and because our powder and goods offered a strong temptation to such a character as Henrick.

As all this was very likely to happen, I let Choubib return to his place with riding oxen, to bring, by a nearer route, his family and property here, to be protected by Kuisip; and next day, dragging the waggon away from the trees on the river's bank, to an open place, and cutting all the acacia and ebony bushes down within one hundred yards of it, I pitched the tent and surrounded it and the waggon with a circular fence, had all the cartridge boxes filled, the muskets new flinted, rockets tied, and all in order for a skirmish.

We lay this way for two days, and kept a very bright look out. I believe we had spies about us, but being ten in number and with plenty of ammunition, we were not attacked, though we were not averse to "play at balls," for I was provoked to chastise, if I could, the miscreant Henrick. Eventually we heard, that in consequence of the conference we had had with him, and his being afraid of a commando, he had sent to the Bath, for Choubib, all that remained *uneaten* of his herd and flock.

Another danger now threatened us—the loss of character with its attendant evils. A stout Bastaard blacksmith, of the name of Martinus, who had been assisting old Arnoldus, (beside whose huts we lay,) having observed our preparations for a skirmish, chose to attribute them to a different cause from the real one, and saddling up his ox, he rode off in the night to Choubib's place, where, with old Dirk and Choubib, he sat down in council, and said, " Depend upon it, this white man has come into the country for no good, he is now preparing to take it from us—will probably go on to the sea and get a commando from a ship and kill us all. You, Choubib, are a great fool for going with him, I would not go a step; and you, Dirk, ought to get back the oxen you lent, immediately."

Though Dirk believed Martinus, and sent for his oxen, yet Choubib fortunately turned a deaf ear to him, and rejoined me with his people, and in a few hours his mat huts were erected by the banks of the Kamop, where it was intended that they should for the present remain.

We had bathed and fished in a deep hole of the river, had shot birds, and had seen the river rise and fill its stoney bed, and noisely run off to the Great Fish river, when the messenger I had dispatched arrived with the good news that the chief Kuisip (roots) was approaching. I went out to meet him, and he advanced towards me; a good-looking, well-grown man, of the age of forty, at the head of a band of followers. I held out my hand to him, which he took laughingly; his people also seemed highly pleased. Perhaps my demeanour was different from what they expected in a white man. I was surprised to find that the chief did not give me the whole of his hand; but I afterwards discovered, that his gun having burst on one occasion, three of his fingers were contracted by the explosion.

I carried Kuisip into my fortified camp, explained why I was so posted, said that now he having come, Henrick would not think of attacking us; and treating him with plenty of tobacco, and giving him a chief's present of a government medal and chain, green cloak, and handsome pipe, I asked him to dinner in the evening. His wife came shortly afterwards, with one or two "maids of honour." The great lady was a good looking and large woman, carrying over her head, to protect her from the heat of the sun, a sort of parasol of ostrich feathers set on a stick. She wore a very handsome kaross of black and red jackal's skins; her half petticoat was tastefully ornamented in front with beads arranged in various patterns, and she walked with an air of conscious superiority, which in a native lady is rather laughable, as it is accompanied by an odd movement behind. I presented her with a blue and white handkerchief for her head, a figured shawl, a striped orange and white petticoat, many strings of beads, and, to complete the favourable impression, I handed her a pipe of tobacco, which she incontinently smoked with great *gusto,* and then passed it to her attendants.

In the evening, Kuisip returned, and I gave him a plentiful mess of boiled mutton, washed down with tea, and then entertained him and the

ladies with a musical snuff box, the fame of which
spread far and wide. Kuisip had with him several
old men, one of whom was an alarmist. "You
will never," he said, "get to the sea: there is
'The Bull's Mouth Pass,' for instance, where
the trees are so thick and so large that it will
take days to cut a way through them—where
lions are running like a flock of sheep—and
where there are Boschmans behind every rock.
Beyond the Bull's Mouth Pass are nations, whose
feet are as broad as those of elephants, and who
are so strong that they can take up a large
ostrich, which is as heavy as a bullock, and carry
it off on their shoulders. Through these people
you cannot pass alive." To all which I answered,
"We shall try."

With stories of the chase the evening was concluded. As a specimen of these tales, it was
said, that during a hunt, a man having quarrelled
with his comrades, made a fire at night, at some
distance from the rest, and sitting at it, he began
to doze with his head resting on his knees; presently, and without his being aware of it, a great
lion came up, sat down opposite to him, and

looked at him across the blaze. The fire began to wax low, and the man, without raising his head, put out his hand and lazily pushed a half-burnt stick further into the fire; the sparks flew out and frightened the lion, which went off with a growl, thus giving the first intimation to the Namaqua of his danger.

Baboons are the best watchmen for lions; and if the king of beasts comes near them at night, they set up such a clamour that he is fain to retire; so that the safest place for a benighted traveller to sleep, is under the rock which the baboons frequent. Though even there, if the baboons are spitefully inclined, he may be between " the devil and the deep sea."

On the 20th February, the chief, according to Namaqua usage, presented me with six sheep, and gave me a grand reed dance, as follows:—
A dozen men assembled, and with reeds, which, closed at one end, were from one foot long to seven, like the horns, of different sizes, of the Russian horn bands, the music of which I used to hear float like that of a grand piano, over the waters of the Neva. Women and girls also came,

and, throwing off their karosses, stood by. One man then blew on his reed, holding it in the left hand, with the fingers opening and shutting to modulate the sound, whilst in his right hand, pressed close to his ear, he held a slight stick to clear the reed; the leader blew strongly, his head stooping forwards, and his feet stamping the ground to beat time; the others blew also, to accompany their leader; wild music arose, whilst the musicians circled round, looking inward, stooping and beating time. The music quickened, the women sang, then sprang forward, clapping their hands, and ran round the circle of reed players, giving their bodies various odd twists, and ending by dexterously throwing up the skirt of their skin half-petticoat behind, previous to falling into their places. Sometimes the women got into the middle, and the men stamped and blew their reeds round them; and thus they continued for two or three hours, with occasional pauses, to favour me with the reed dance, which I had never seen or heard of before.

A present of tobacco rewarded the musicians;

and now I had so much won upon Kuisip, that he declared, if I wished his services, he would accompany me to the sea with twelve men, and only asked for a musket for himself, and some small articles for the escort. I readily agreed to Kuisip's terms, and was glad that he had proposed to go with me through a country, as the phrase was, " sharp with wild men and beasts."

On the 22nd we set out from Nanebis, after I had given Kuisip's and Choubib's people some powder and bullets to defend themselves with, in case Henrick should take it into his head to pay them a visit. Travelling in a westerly direction for some distance, along the left bank of the Kamop; we crossed it, and outspanning during the heat of the day, we shot some rare and beautiful black backed thrushes with crimson breasts; three and a half hours more brought us to the dry bed of a river, where we slept.

Next day we crossed the Koahap (coming on), which flows from the Gnutuas (black morass), westward, and four and a half hours more brought us to the 'Oup, or Fish river, here eighty yards wide, flowing briskly in a southerly direction, and between high banks fringed with wood, to the Orange river.

The 'Oup was a very pleasant sight for us; it is the chief drain for Namaqua land, rises about the tropic, and two or three dozen rivers empty themselves into it from the east and west. There is also a waterfall of some height, about the middle of its course. The Fish river is not always flowing, but in its bed there is constantly plenty of water in large pools, or sea-cow holes. Though much inferior to the Orange in size and beauty, yet the 'Oup is an important river, is possessed of considerable variety of scenery, abounds in strange birds along its course, to reward a collector, and the fish are similar to those found in the Orange river.

The heat, for some days past, had been, at mid day, either 105° or 110°, and when we crossed the Fish river and outspanned above the highest water marks, we gladly plunged into the refreshing stream, and then fished with hooks, by moonlight. Though it is not proper to eat or drink much at any time, and particularly in hot

weather, yet the temperature to which we had
been lately subjected, caused my people to forget
all prudential considerations, and when they now
got a bambus of water to their heads, they did
not leave off before they had got half a gallon at
least under their belt. Like the celebrated
Abernethy with his appetite, " they had such a
devil of a thirst they could not help indulging it."

The heat again made the people irritable, and
there were some angry disputes between Choubib,
(whose temper was none of the best,) and my
men, and also between the men themselves. I
had some difficulty in keeping all quiet. After
breakfast, on the 24th, I stripped, and, with my
people, we hauled the seine, but we were not
so successful with the net as with hooks.
The occupation however was agreeable, and it
prevented squabbling.

The Boschmans have a peculiar mode of fishing in the 'Oup river; they make conical baskets
of stick grass, which is as thick and hard as quills;
some of them then tie a stone to the back of their
necks, to keep them down in the water; and,
wading in, they sit down in the river with the

water up to their mouths, and the basket between their legs, the mouth of it to the front; other Boschmans wade so as to drive the fish towards the basket-men, who are sitting in line, and who, pushing the passing fish into their baskets, collect a number in them, then rising rapidly, they empty them on the bank, where sit their women, and then resume their place in the river.

To free ourselves from the rugged hills which enclose the 'Oup, we travelled west, by passes which would have been impracticable for any other vehicle than a long and elastic Cape waggon. We ascended steep acclivities, descended over rocks and stones, and then traversed the sides of hills where we were compelled to hold up the waggon with long riems or straps of leather, to which a dozen men clung, to prevent the waggon going over on one side. It had several narrow escapes; and if it had fallen, it must have been dashed to pieces, and carried a valuable span of oxen with it.

After five hours hard work, but during which time we had barely accomplished ten miles, we reached a wooded dell (under cliffs) called Neims (giraffe). Here a little water was painfully collected, and we felt the night as hot as the day, so that the people slept perfectly naked. Thermometer, at sunrise, 85°.

The hills about the Fish river abound in klip springer bucks, or rock leapers, which, like the chamois of Switzerland, are found in the most inaccessible places. The length of the klip springer is three feet two or three inches; its height twenty-two inches; the colour of its back is greenish yellow; the belly is ferruginous. The hair of the body is very peculiar, being like thick mossy grass: it is easily crushed between the fingers, and wants the usual elasticity of hair: it is in great request in the Cape Colony for saddle-pads. The head is short and compact; the horns of the male short, straight, and sharp; the tail is hardly to be distinguished; the legs are very robust; the pastern joints are stiff, and the hoofs round; all which enable the animal to display the greatest agility among its favourite rocks.

The hunter who attempts to follow the klip

springer must share in its activity. Lightly
equipped, and assisted by his dogs, he must drive
it before him over the broken and bushy sides of
the hills, towards peaks, which it scales in a
most surprising manner, and where, on attaining
the summits, with the dogs helplessly barking
and whining below, it fancies itself in perfect
security; but the proper distance being attained
by the hunter for his rifle to tell, a whistling
bullet drives the once nimble buck headlong
down the steep, to afford the reward of its delicious venison.

Leaving the hills, we traversed bushy plains,
and outspanned at night at a dry place. I had
a small supply of water for my own people, and
I imagined that in this thirsty land the Namaquas would never have gone any distance from
home without skins for holding water, as is the
custom in the East; but a small supply in a
bladder, stomach of a buck, or ostrich shell, is
all they commonly have—and not always that;
and he is considered to be extraordinarily pro-
vident who skins a goat by the neck, so as to
make a water sack.

Kiusip's people got their allowance of flesh, but said they could not eat it unless I spared them a little water. I gave them a panikin each, and not knowing when we should meet with water again, I lay down under the small barrel, which was attached to the rear of the waggon, and slept there, to prevent its being tapped during the night. The thermometer had been at 105° during the day; we were all very thirsty, and I dreamt of rushing streams.

Next day, steering S.S.W. (!) for water, our progress was over an extensive plain; and in the distance was a line of high trees, for which we made: it was the course of the periodical river Nukanip, (black bulb).

We outspanned at a place called Habunap, (devouring) where we found the "lay places" of two or three Boschman families. They consisted of mere hollows scooped in the sand (like ostrich nests) under the bushes, and were strewed with dry leaves. The habits of these people are more abominable, and they seem to have less regard to cleanliness than dogs: they had fled on our approach. This small community, we were told, lived on fish from the 'Oup river, and on gum, ants, and bulbs, according to the season of the year.

The poor oxen got no water here, owing to their having been sent off to a reported water place down the river, but where they found none; we dug in the sand at the outspan, and made a water place; but, in the mean time, the oxen all ran off, and next morning there was not one to be seen.

The people went in various directions in search of the oxen. Two of my men, Robert and Henrick, among the rocky glens of the Fish River, were thirty-six hours without food. They saw plenty of water, and of klip springers, and other bucks, but could not get a shot at any of them, and at night did not light a fire for fear of the Boschmans.

We were three days without the oxen! and were completely at a stand' still. My feelings during this time were not to be envied, as I thought the cattle had been carried off. At mid day the heat was 105°. In the evenings I went out with my gun, and saw plenty of guinea-fowl, nearly similar to those so well known in England; but the cocks were larger, and were very fine birds indeed. Elliott shot here two very small hawks, with speckled plumage, and which had been feeding on lizards: these hawks, if not new, are at any rate, exceedingly rare.

The oxen came back in three clumps from the Fish River (thirty-five miles off). The largest was brought single-handed by Kewit, an active old padwijzer, or guide, who smoked a straight Namaqua stone pipe, which could hold a week's supply of tobacco. I charged this to the muzzle when I saw him again, and gladdened his heart with many sticks of the precious weed, besides adding thereto a couple of knives and of tinder-boxes, and he was now considered a very rich man!—the value of his present was little more than half-a-crown!

Again on the 3d of March we were "progressing," and made a road for the waggon by removing the stones from a rocky descent, whilst we journeyed in a north-westerly direction. We passed many camel-thorn trees in a sandy bottom, where the waggon stuck fast repeatedly, and we worked hard with the spades to dig it out, and pushed at the wheels to assist the bullocks.

The camel-thorn, or acacia giraffæ, deriving its name from the fondness of the cameleopards for its leaves, is a lofty tree, whose foliage is disposed from the top downwards in umbrella-shaped masses; its wood is hard, heavy, and of a deep red colour; the bark is very coarse; its pods are oval and solid; and the thorns are thick and brown. We met with it in the open plains, at a distance from water, and often derived comfort from its shade, though many of the sun's rays penetrated between the light and fringe-like leaves.

> " Oh ! Abyssynian tree !
>   How the traveller blesses thee,
> When the night no moon allows,
>   And the sun-set hour is near,
> And thou bendest thy boughs
> To kiss his brows—
>   Saying " Come rest thee here !"
>   Oh ! Abyssinian tree !
>   Thus bow thy head to me !"

I now heard, for the first time, that we could

expect no water for the oxen for three days, and I could not help, therefore, giving way to melancholy reflections. *Cui bono ?*—of what use is it? I thought, traversing the wilderness in search of a land of promise to the north, if we are to be constantly losing the cattle, and probably perishing from thirst ourselves? Is it likely that this expedition will repay me for all the fatigue and hardship attending it, and will it be appreciated in England? We fear that this travelling may be, like virtue, "its own reward." Still we must not despond; it is sinful to do so. We must struggle on, *à ma puissance*, to the best of our ability. "Surely, whispered an under-current of hope, "your face shall not be blackened, nor shall it be said that you have now journeyed in vain. Place yourself then *au plaisir fort de Dieu* —at the Almighty disposal of God.

Sitting on a rock one day, thinking of the unknown land beyond me, of home and all its enjoyments, something tickled my naked ancle, and brushing it with my hand, a large hairy spider bit me with its poisonous forceps, and then disappeared in a crevice. The ankle swelled very much at the time, and now pained

me so much as to occasion lameness, and yet I
could not "strike work."

We crossed the Kusis, (close by) a branch of
the Koanquip, and which is again a branch of
the Great Fish river; but saw no water, and the
heat was very great. Those only who have experienced the tongue cleaving to the roof of the
mouth from thirst, and the lips ready to crack,
can know what we now experienced. We were
all this time moving over sand, and between
bare hills, and were quite uncertain as to finding
water at our outspan. I tried to say something
to comfort my followers, but " the flesh was
weak," and I could not utter a word, and we
passed on in gloomy silence; but a change soon
came over the party.

The word was passed that zebras were in sight,
and immediately every one was alive. In the
plain, a troop of half a dozen zebras was seen
grazing; they were unaware of the approach of
danger, till the people had rapidly extended
themselves, so as to cut them off; the zebras
then took the alarm, and headed by a noble
stallion, they galloped off with a cloud of dust
behind them toward a pass; but here they were

turned by a detachment of pack-oxen; they then took to the hill side, but a hunter running towards a bush, and lying down out of sight, fired as they stood for a moment near him to gaze, for they seldom run far without turning to look round them. The ball took effect on a fine mare, and she fell with the blood gushing from a mortal wound in her head; the others disappeared, dust and stones flying behind them. Two oxen were driven to the dead zebra, she was soon cut up and packed on them, also a beautiful unborn foal.

As we approached the outspan, the sheep began to smell the air, then commenced bleating and running, followed by the oxen: when the whole throwing themselves into the bed of the Kusis, they filled a shallow pool of green water, from which they eagerly drunk, stirring it up all the while into liquid mud, and then showed the bestial nature by polluting it terribly; but, notwithstanding all this, the people drunk greedily among the feet of the beasts, and thus afforded another instance of what one must put up with in an African journey.

By digging, we got a good supply of water under the sandy bed of the river, and we also found abundance of prickly cucumbers, green, with white stripes. One of the men brought me eight young ostriches, which he had just taken near a nest. They were the strangest and most old fashioned looking little things that can be imagined; with black and brown spotted heads, large orifices of the ears, black and white moss-like hair on the back, short wings, and legs like pillars of dark skin, they looked abroad always towards the wilderness, repeating a low churr! churr! for their mother. I did all I could to preserve them, but they were all dead in a week, one after another.

We halted at the Kusis on Sunday to refresh the poor cattle, and found it with abundant wood, grass, and water, an agreeable outspan, thus having realized the saying,

" The help is nighest
When need is highest."

On Monday, we journeyed on smoothly for seven hours, and at night reached the long deserted station of Bethany.

CHAPTER X.

Sketch of the deserted Station of Bethany—How it came to be abandoned — A Skirmish — Cool Determination of a Namaqua—The Author is well received by the Tribe of Buys — Cross the Koanquip River —Arrive at Nanees— Jan Buys—A Word on Smoking — Beautiful Landscapes —The Lion and the Zebra—Wild Flesh—The Black Lion —Tuais—The 'Un'uma Mountains—Henrick the Hunter —His extraordinary Fleetness—Festival at Tuais—Offer of Marriage—A Tale of Jealousy—The Strength of the Expedition is increased—Alarm of a Lion—Leave Tuais— Another Conquest—Kill a new Fox—A Marriage Gift— Mount D'Urban — The Game of Hous — Henrick Buys gives a Sample of his Ability—Following the Spoor—Bee Hunting—The Borders of Boschman land.

THE mission station of Bethany was established by the Rev. Mr. Schmelen, far in the wilds

of Great Namaqua land, and here at this distant point from civilized society, he endeavoured
for some years to propagate the truths of the
Christian religion among a heathen people. The
site was selected from its possessing one of the
most abundant fountains in the land. The water,
in a stream sufficient to turn a mill, issues clear
and sparkling from a low ridge of limestone, in
the midst of a great plain, and irrigates a large
field, once divided into gardens, and now exhibiting only the remains of former cultivation.

The ruins of the station stand on the same
ridge from which the waters of the spring issue,
and consist of the roofless walls and gables of a
single storied dwelling-house, of a long and low
chapel, and of outbuildings, all of which had
suffered by fire. They command an extensive
view of the plain to the east, bounded by a varied
range of mountains. Trees are not wanting, and
some distance in the rear are the wooded banks
of the Koanquip, or " off running " river. Altogether, the site of the station is pleasant and
agreeable to the eye, and it is only eight days
journey from the whaling bay of Angra Piquena.

How the station came to be abandoned is thus told. There lived under the charge of Mr. Schmelen the chief Amral and his people, the families of the Buys, and the old Namaquas under Habusomop. Amral and the head of the Buys had been in the colony, were therefore called " overlams," and were cleverer than the Namaquas of Habusomop. The latter were jealous of the superiority of the former; besides, they wanted to get the whole station to themselves, and they long devised the means to get up a quarrel. At last they took the opportunity of Mr. Schmelen's absence, on one occasion, to send a scoundrel of their party into the missionary's sheepfold, to break the legs of a number of his sheep. When Mr. Schmelen returned, he was surprised and amazed at the loss he had sustained, sent among the people to inquire who had done the mischief, and asked the head people to assemble to investigate the matter. Habusomop and his people fearing detection, and that their agent would inform against them, killed him privately, and collecting their guns and other weapons in one of the huts over-night,

Habusomop came next morning with a bold front to the council of elders, before the mission house.

Mr. Schmelen left the case in the hands of Amral, and did not interfere. The investigation was proceeding, when angry words ensuing, Habusomop took up a stone and threw it at old Buys, when the whole rose up under strong excitement, and Amral seeing how matters were tending, called out to his people and to the Buys to run for their arms. Bullets and stones soon began to fly. The people of Habusomop being ten times the number of the Overlams, and besides being all ready for a fight, had the advantage at first, but they did not enjoy it long. The Overlams united and drove the old Namaquas before them off the station, and over the plain like sheep. Half a dozen were killed and wounded on both sides. The second son of old Buys, Henrick, the fleetest runner in Namaqua land, got hemmed in among the Habusomops, and got out alive by stooping and by his agility, but received a ball through his left arm. Choubib, my interpreter, was knocked down by a blow on the head; and his brother, after one discharge from a double-barrelled piece, fell with a mortal wound, but still retained his consciousness. One of the Habusomops ran up to him, and thinking he was unloaded, he seized the gun by the muzzle, but immediately received the contents of the second barrel through his heart. Choubib's brother then taking the bullets out of his own pouch, threw them from him to a distance, and opening his powder-horn, scattered its contents among the sand where he lay, that his ammunition might not be used by his enemies. He then fell back and expired, exhibiting a remarkable instance of cool determination in his last moments.

For a week after this skirmish, the Habusomops used to come at night and fire dropping shots at the mission house, and into the huts of the Overlams; and they also shot some cattle. Several Bastaards at the Great River hearing how their esteemed friend Mr. Schmelen was situated, came, headed by old Wm. Joseph, my "Copper-indicator," to fetch him away. Mr. Schmelen seeing no end to the bad blood which had been engendered at Bethany, and having for sometime tried in vain to prevent the people of the station exchanging their cattle at the sea, for fire-arms and ammunition, now quitted Bethany and retired to Komakas, within the colony. The Overlams also quitted Bethany for a time, when Habusomop set fire to the buildings. Amral and his people went and lived east of Bethany, and beyond the Great Fish river, where they now are; and the Buys returned and took possession of the station, where we found several families of them in mat huts round the ruins.

Old Buys, the founder of the tribe, had been long dead, but he had left many sons, all like himself great hunters, and who had all families. Some of the younger sons were at Bethany, and we found them very civil, though we were led to believe by the Bondelzwarts that we should be undoubtedly robbed and murdered by the Buys. So far, however, was this from being the case, that, as shall afterwards be shown, we must have perished if it had not been for the great assistance we received from the Buys!

Shop was opened again, some more oxen and

sheep were bartered for our goods, and the dance was kept up till a late hour at night: this, with a few pipes of tobacco, put all hands in high glee.

I remained two days at Bethany, and left on the 9th of March for Jan Buys (the eldest of the family's) place. We travelled in a northerly direction over a plain, and after twenty-six miles journey, we outspanned on the banks of the Koanquip river, in which were pools of water. Next morning we crossed the river, with great difficulty, from the reluctance of the young oxen, our new purchases, to drag the waggon; they broke the yokes, kicked violently, pushed with their horns, and lay down. We went on for a short distance, and then halted at the Tamu-hap (or dropping) river, opposite the Kurusap, or Sour hill. Here a beautiful span of red oxen arrived from Jan Buys, to help us to his place, at which we arrived after three hours journey.

Nanees, (corner) where Jan Buys lay, has a pleasant and elevated site among the interesting group of the Tamuhap hills. There were six huts on a grassy plateau, and plenty of cattle and sheep at Nanees.   Rain fell, and the temperature was now too cold for us, 70° at midday.   Jan Buys dined, drank tea, and smoked with me in the evening; that is, smoked by himself; for remembering the delicious hukas of India, the kaleuns of Persia, and the chebuks of Turkey, I have not touched tobacco for ten years. I say it, and without affectation, that the European manner of smoking is coarse and vile, and to me disgusting.   We all know it is so to ladies, and surely the aspirants for lady's favour ought to lay it aside, or else adopt the oriental refinements of water-pipes and purified tobacco.

But enough of this.   Let us talk of old Jan, as he was commonly called, though apparently he was not more than five and forty; but he was stout and lusty, and the oldest of the brothers, " the old" being a term of respect and regard. Jan was reserved at first, but under the influence of the moderate excitement of tea and tobacco, for I had no spirits, he soon opened out, and I had more agreeable conversation with him in Dutch, which he had learnt from his father, than I had had with any one else for a long time.

He offered to go with me to his brother Henrick the hunter's place, in advance, and there to make arrangements for assisting us onwards.

Again inspanning, we descended to a most beautiful plain, which waved with high white grass, like a harvest ripe for the sickle; there were numerous ant heaps among this, a sign of abundant pasture; clumps of green mimosa, and brown and deep red hills, three or four hundred feet high: altogether the scene was most delightful, and I highly enjoyed it, with the agreeable temperature of 75° at mid-day.

Game was not wanting. We scattered ourselves over the country. The traces of zebras were seen, and then the fresh print of a lion's paw; presently the king of beasts was observed at a distance devouring a zebra under a bush; having had his belly full, he slowly moved off, and left half of his prey, which was immediately packed on bullocks and carried off for supper.

Though it might appear very sporting and Nimrodish to say that I relish wild horse flesh, with its yellow fat, yet I cannot bear it if I can get anything else. It is true that I can eat meat,

" done or under done, raw or cooked," when hungry, and so can most men; but to say that I preferred coarse and rank food, which wild horse is, to good beef and mutton, would be (to use a vulgar expression) humbug!

Thinking that the lion would pay a visit to our larder in the night, my men slept on the sand beside me, with swords and bayonets fixed on their rifles and fusees, after setting fire to some old nests of republican birds in the trees, to scare our expected visitant; but he did not trouble us.

" One day," said Jan Buys, " I was in the field with my gun, looking for game, and I saw before me, beside a bush, a large black body. Thinking that it was an ostrich resting its long neck and head on the ground, I approached quietly to shoot it. It rose without seeing me, and I discovered it to be a long haired black lion. I concealed myself as well as I could, and watched its motions. It stretched itself, looked round, and then came towards where I lay, at a trot, its mouth open, its horrid teeth as long as my fingers, and grumbling to itself from hunger.

I thought my last evening was come, for I never before had seen a lion so shaggy, so black or terrible as this was. I took aim at him with my gun, through the bush, and was going to fire as he was passing me, when I suddenly remembered a warning our father, old Cobus, gave us, when he showed his sons the proper manner of hunting. ' Never fire at a lion,' he said, ' when you see he is on the move by daylight, for he is then very hungry, and looking after food; and if you miss or only wound him, you are dead men.' Accordingly I ran off as hard as I could, without firing, to a hill where my people were, and left the black lion to continue his hunt."

On the 13th of March we continued our journey over the same delightful country as we had seen the day before, and in two hours and a half descended to the bank of the Gnuanuip river (whose name is untranslatable,) and outspanned at Tuais (or Mud) Fountain, where lay Henrick *the hunter*, " par excellence."

From Tuais we saw the long line of the 'Un'uma, or Bulb mountains, two or three thousand feet high, east of us, and between us and them was the Koanquip, a branch of which was the Gnuanuip.

In the evening, Henrick came from his huts to visit me; he was a spare made and athletic Namaqua, of forty years of age, about five feet eight inches in height, nose low, but inclined to aquiline, teeth rather prominent, but covered with his lips; a good-humoured smile about his mouth, and altogether with a very amiable and intelligent expression of countenance. He was beautifully formed, deep chested, small waist, and muscular arms, thighs, and calves, without any extra flesh beyond what was necessary to give perfect symetry to his figure. His feet were small, as is usual among the Namaquas, but his instep was particularly high, and even rose in a sort of knob in the middle: this may have added to his astonishing power as a runner.

The reader must not imagine I indulge in a traveller's license when I say that Henrick could catch and kill zebras by fleetness of foot: I believe he has often done this; for I have seen him turn zebras towards the guns; and when I tell how he managed to catch them, I may be believed; if not, I must lie under an evil imputation, which I would willingly avoid—for, *Hora et sempre*, now and always " Truth me guide."

When Henrick's powder ran short he took a hunting knife in his left hand (for he was left handed, and continued so, though it was through his left arm he had received a ball at the skirmish at Bethany) and knowing there were zebras in his neighbourhood, he went out to the field to seek them, to their grazing ground, and to run them down.

Walking on his toes with an elastic springing step, at the rate of upwards of five miles an hour; he paced over the plain glancing at the ground for foot marks, and on each side of him, with his keen eyes. The prints of the compact hoofed zebras are observed on the sands, and presently the troop itself is seen grazing near. Henrick stoops, disencumbers himself of every skin covering which might encumber him, even to his leopard skin cap, and steals as near as he can to the game without being perceived; but the watchful eye of the stallion discovers the hunter, when he leaves the cover of the bushes, and giving the

alarm to the rest, the whole gallop off.   Henrick,
without putting himself to the top of his speed
at first, follows them; the zebras stop to graze,
Henrick running like a race horse, with his stomach near to the ground, bounds toward them.
Away they rush again, snorting, and tossing
their striped heads in the air, and switching their
light and mule-like tails in the pride of fancied
fleetness and freedom.   The hunter "sits on
their skirts," and relaxes not from his pursuit for a
moment; he clears stones, bushes, and other impediments; after three or four miles he is in perfect
wind; the ground seems to fly from under him;
and, as he expressed it, he was now unable to
distinguish the heaven from the earth.   The
zebras stop and gaze occasionally, as before;
but it is now but for an instant, for their enemy
is closing with them; he drives them towards a
steep face of rock; they hesitate about the means
of escape; Henrick is amongst them in a moment, and seizing one of the striped troop by
the tail, he swings it to one side, throwing the
whole weight of his own body towards the ground
at the same time.   The zebra falls on its side,

when Henrick instantly plunges his knife into its chest, and then allows it to rise and run off; it keeps up with the rest for a short distance, then gradually falls behind the troop, weak from loss of blood. Its comrades wait for it till Henrick again dashes forward, repeats his fatal thrust, and if two stabs are not sufficient to stretch the zebra dead on the plain, a third is given, which rolls the beautiful body lifeless on the ground, covered with dust and perspiration. The successful hunter then returns to his huts to send his people with pack oxen to bring home the prize.

Henrick is rivalled now in fleetness by his eldest son, Jan, which would not be, says the father, if it were not for his own crippled arm. Lately, the two were out in pursuit of giraffes, and getting close to three, the father told his son to assist him in stabbing the last; but Jan said, " No : let us attack the first and largest." Which they did; and after a smart pursuit, managed to stab the first with fatal effect.

It must be borne in mind that horses come up with both zebras and giraffes, but still the powers of both Henrick Buys and his son must be very astonishing to enable them to rival horses, and thus to show what a man is capable of accomplishing with temperance and training.

On the 14th March it thundered and rained all day, to our great refreshment. Thermometer at sunrise 60°, at mid day 75°. I gave presents, as usual, to Jan and Henrick Buys, and to their wives, whilst tobacco and the fiddle put every one in good spirits; and on the advent of the expedition to Tuais, there occurred a sort of festival among the dependants of the two brothers. Some of the dancers showed a good deal of humour in their performances of imitation reels — they stooped near the ground in going through the figure, then sprang up, beat the ground rapidly with their heels, kicked up before and behind, stopped suddenly, looked their partner earnestly in the face, and with the cry of "hoet," bobbed their heads forward, and then went through the figure singing an accompaniment. There was a little Damara too, with a pot belly, who imitated the reed and fiddle very comically, accompanied with dancing. There were exhibitions with bows and arrows, lances, and wrestling, &c. Presents of cows and sheep were brought, but which, of course, I took care to pay for. More sheep were also bought for cotton handkerchiefs, which were here so wonderfully attractive that one of the prettiest girls in the place, or that I had seen on the north side of the Orange river, came, and, to my surprise, offered to marry me for a cotton handkerchief! I was thus in the predicament of St. Anthony in the Wilderness.

The Namaquas are sometimes troubled with jealousy as well as other people, and cherish their revenge, like barbarians in general, till the fitting moment for gratifying it. Thus it happened, that two young men made love to one girl, but, as is customary, she attached herself to one of them only. "The Rejected" did not forget or forgive the affront, and for thirty years he looked out continually for an opportunity to destroy his rival. Like Shylock, his revenge was implacable.

"I'll have my bond; I will not hear thee speak;
I'll have my bond; and therefore speak no more.
I'll not be made a soft and dull-eyed fool,
To shake the head, relent, and sigh, and yield
To intercessors."

At last it happened that the two (now about fifty years of age) went out together alone, on a honey hunt. The bees having collected their treasures on the face of a cliff, at some distance below its summit, it became necessary for one of the men to be lowered to collect the spoil.

" The Rejected" lowered his ancient rival by riems, or thongs of leather, and when a skin bag was filled, he hauled him up again to the brink of the precipice, and then took the bag from him. " Is that all ?" he coolly inquired. " Yes," was answered: and now Hell was at work. " You cut me off from the woman I wanted," cried the demon, "and now I cut you off." With that he divided the riems (which were made fast to a bush) with his knife, and his screaming victim was precipitated to the bottom of the cliff, and dashed to pieces !

The murderer returned to the huts; the dead man's wife asked where her husband was; and she was answered, by the murderer, that he did not know, for that they had gone different ways in the field; but the spoor, or foot marks, having been " taken up," the mangled body was soon discovered. However, so little law is there in this *happy* land, that the murderer was allowed to live, and he comforted himself not only with his own wife, but with his first love also, who now, poor creature, could not bear being alone in the world!

I invited Henrick to be my chief hunter, and he agreed to go with me. I asked Jan to assist me with oxen for the waggon; he said he should do so, and not only that, but he would accompany me himself to the sea with some men, for there was danger on the road, and the party required to be numerous. I thought my party of twenty-one strong enough already, and more manageable as to food, but as the Buys had been so kind, and as I had been very fortunate to get assistance from them so readily—from those men whom I was told to look on as cutthroats and robbers—I could not object to their taking what men they choose; and my party, with some boys, became fifty in number: in fact, a considerable commando with four headmen, Kuisip, Jan, Henrick, and Choubib.

At night, on the 16th, there rose a cry, that there was a lion seen passing the huts on the bank of the river: the Namaquas lying round me behind two or three half-circles of bushes, rose in a moment, seized their arms, and calling to one another, they rushed to the huts like an angry swarm of bees. As the night was dark, and the Namaquas had evidently more the intention of frightening the lion away than of getting a shot at him, I kept my own men prepared for a visit from the enemy, who was looking after our cattle, which had been quietly reposing, like ourselves, under the trees. But having heard the noise of the people, he, the lion, I suppose " thought better of it," and went off without troubling us more, and therefore, though I wanted a hunt next day, the Namaquas said, as usual, " since he did no harm, we must not go after him."

On the 18th, we left Tuais and travelled northward over the same beautiful grassy plains as before. I was not the only one who made a conquest at Tuais, or rather that a handkerchief subdued (for I dont pretend to flatter myself). Another stout Namaqua girl, on this, the morning of our departure, appeared with kaross tied up for walking, horn for fat, and tortoise shells full of buku attached to her girdle, and all prepared to follow the fortunes of Robert; and I had some difficulty in persuading her to remain behind, as our journey was rough enough for men; women it must have killed; besides, decency forbade our travelling attended with the softer sex. I had made Henrick the driver leave his helpmate at Lily Fountain, though she was our cook and washerwoman.

We had plenty of excitement the first march from Tuais; we saw ostriches, steenboks, hawks, owls, black and white crows, hares, tortoises, scorpions; pursued, but in vain, a sort of cat with a red skin; chased a beautiful orange-coloured martin with a long tail, into a hole, and tried to dig it out (without success), thinking it was a novel species; and I caught, with the assistance of the dogs, a fox of a species which Choubib and the others said was new to them; its colour was grey; muzzle, face, ears, ridge of back, legs, and half of the tail, black. It was two feet seven inches long including the bushy tail. After seven hours inspan we halted

for the night on the banks of the Great Koanquip river.

I had collected fourteen black and red jackals'
skins of the most beautiful fur in South Africa
to make a cloak of, intending it for whoever I
might eventually induce to enter into a " holy
alliance." The making up of this marriage present was interesting. The skins, after having
been well stretched and scraped, were carried
about by some of the Namaquas and rubbed
with the hand, when they had nothing better to
do, and became, by this mechanical process, as
soft as glove leather; they were then put under
ground and covered with moist sand for a short
time to flatten them out; then the edges were
cut straight, and the seven skins of the upper
row retained their heads and ears, whilst those
of the lower were neatly scolloped at the bottom;
two broad flaps were set apart for the upper
corners to fold over the breast. The whole were
then exceedingly well sewn together with sinews,
and formed a very handsome fur cloak, attached
to the person with double thongs at the neck.
How this cloak was disposed of, I have already

noticed, and I have therefore no occasion now to say, (as a distinguished writer of the day said when he hinted a wish to be adopted by any one) " every necessary information will be given at the respectable publishers."

We again advanced in a N.E. direction, and passing under a bold mass of mountain about two thousand feet high, with a square top, rendered apparently inaccessible by scarped cliffs of red sand stone in horizontal strata round it, and having attached to it a conical hill south of it, we halted for some hours on the wooded banks of the Karee or Little Koanquip. I named the mountain after the excellent Governor of the Colony of the Cape of Good Hope, Mount D'Urban.

The Namaquas here played in the sand a very noisy game called " Hous," and which they now constantly practised at every convenient opportunity. Three or four men sat down on the sand opposite to three or four others, with four rows of holes, twenty-six in each row, between them, into each of which holes two or three seeds were distributed, and changed about from one hole to the other accompanied with shouts, screams, and vociferations the most extraordinary—although they played for nothing.

Henrick Buys now gave us a proof of his ability: he went in front with his gun, crept upon a herd of springboks, and wounding a fine young male, " in pride of grease," through the rear hind thigh, he went after it like the wind, whilst it followed the herd rapidly on three legs; but Henrick getting within a few yards of it, knocked it down with a stone, and we ate its venison for our supper. He also pursued a zebra for two miles; but missing a screw of his gun, he was obliged to halt and retrace his own foot-marks to look for it. As to tracing foot-marks, the Namaquas, like other African hunters, can track a man or beast by marks on the sand, among stones, or by bushes, which would be perfectly unnoticed by a white man's eye. Thus a pebble wanting the dew in the morning shews that it has been turned during the night; and leaves or twigs disturbed, shew a passer by; some marks, as on sand, are available for many days if no rain falls in the meantime.

Bee hunting was very curious. Whilst I was engaged in the chase one day on foot with a Namaqua attendant, he picked up a small stone, looked at it earnestly, then over the plain, and threw it down again, I asked what it was; he said there was the mark of a bee on it; taking it up, I also saw on it a small pointed drop of wax, which had fallen from a bee in its flight. The Namaqua noticed the direction the point of the drop indicated, and walking on, he picked up another stone, also with a drop of wax on it, and so on at considerable intervals, till getting below a crag he looked up, and bees were seen flying across the sky, and in and out of a cleft in the face of the rock. Here of course was the honey that he was in pursuit of. A dry bush is selected, fire is made, the cliff is ascended, and the nest is robbed in the smoke.

In the afternoon of the 19th, we ascended a dangerous and rocky pass, and after two hours journey we halted for the night behind a line of bushes. Here we left behind us Great Na-

maqua land proper, were on the confines of
the country of the Boschmans, whilst the ex-
treme limit south of the negro Damaras of the
Hills was on our left on the table land of the
Great 'Un'uma range.

BULL'S MOUTH PASS.

Etched by William Heath

Published by Henry Colburn, Great Marlborough Street, 1831

CHAPTER XI.

Difficulty of determining the Limits of Tribes in South Africa
—First March in the Country of the Boschmans—Zebra
shooting—The young Gemsbok—Goats—Delicacy of the
Lion—Anecdote—First Rencontre with the Boschmans—
How they eat, creep on Game, frighten Lions, and make
Fire—The Caravan en Route—Musquitoes—More Bosch-
mans—The Women described—A Discovery—What a
Boschman likes best—Boschmans live by Lions—The
Great Fountain—Delicacies—A disgraceful Occurrence—
The Great Flat—The Huntop River—Frog Soup—How
we commonly passed the Night—A Storm—The Narop
River—Locusts—Giraffes—The Mirage—The Bull's
Mouth Pass—A grand Hunt—Author is charged by two
Rhinoceroses.

In South Africa it is impossible to define
with accuracy the exact boundaries of particular
nations or tribes. Nomade people, who are continually shifting their positions or haunts, are
found on the banks of rivers and by fountains
this year, whilst on the next visit to the same
localities, part of an entirely different tribe may
be found there. Thus the country we were now
about to enter, though occasionally visited by

the Great Namaquas, was principally traversed
by Boschmans of the same general appearance
and language as the Namaquas, but darker, and
not so well dressed, and possessing neither flocks
nor herds. They are also exceedingly wild, and
it is necessary to keep always a very good look-
out that they do not steal the cattle, or murder
those who pass through their broad plains, or
approach their craggy mountains.

On the 20th of March we travelled six hours
and a half in the country of the Boschmans
without seeing any signs of inhabitants. Extensive plains were before us; on our left hand,
beyond them to the west, were blue ridges in the
distance between us and the ocean; whilst on
our right the table-topped range of the 'Un'uma
continued. We crossed the Humabib (root
water), where there was grass, water, and shade,
and then missing Henrick the hunter, we looked
out before us, and saw dust rising at the foot of
the mountains. We knew that this must proceed
from large game; accordingly, I rode with Kiusip and his men towards the right, whilst the
rest of the people distributed themselves to cut

off the wild animals, whatever they might be.
We dismounted and lay down behind bushes in
an extended line. The cloud of dust approached
us, and below it seemed to dance several black
bodies; it came near, and seven zebras, following each other, galloped towards us. It now
appeared that Henrick and two or three men
had ran on before, and had turned the zebras
towards our guns. It was a fine sight to witness
the mares, young males, and a foal halt, whilst a
powerful stallion, with his mane as if newly
hogged, and his tail switching his striped thighs,
come on singly to reconnoitre my horse and the
riding oxen. He came close to the bush where
I lay, but I kept quiet till the whole troop should
pass. In the mean time an impatient hunter
crept towards the halted group of plump and
round females, and fired into them without
effect. The stallion snorting wildly, galloped
back to his charge, and the whole passed us
rapidly, and received our discharge; long tracks
of dust followed their heels. One fell on the
plain, as it was breaking the line of the people
facing the mountains, but recovering itself,

it was tracked by its bloody spoor towards the Koanquip, and was nearly come up with by two of the people, when a party of Boschmans appeared, intercepted and secured the prize.

We saw white-rumped springboks ahead, and went after them, and then among hills on the right a troop of gemsboks appeared, the antagonists of lions, with their long straight horns, like that of the fabled unicorn. We made a dash at the gemsboks, and intercepted them as before, between our extended hunters and the mountain; but it fell dusk. The old ones all escaped with apparently slight wounds, whilst only a young fawn, in appearance very like a brown calf, with a short tail, was secured alive. We carried it with us to the waggon, and gave it the cow's milk.

To an African traveller goats are invaluable; they can accompany him every where, and live where cattle would pine away and die; though the cattle of South Africa are assuredly the most hardy and enduring in the world. What care they for clover fields, and meadows of thick green grass, as long as they can range among the

thorn bushes, and fearless of the prickles, get
their tongue round the blades of white and sweet
grass which grows up among the twigs. They
also eat the tops of many bushes, but not of so
many as the goats, each of which give nearly as
much milk as a Namaqua cow, and of much
richer quality. The milk I obtained from my
four or five milch goats (though at times, when
we were hard pressed for water, it only amounted
to a pint morning and evening amongst them),
was a source of great comfort and support.

Next day, we went on again, and passed
a place, where two or three years before, some
of the people had found a cameleopard, dead
and entire, though it had been just killed by a
lion, which had climbed from behind and struck
it down. The people accounted for the lion
having refused the flesh of the giraffe, because
in its fall the contents of its stomach had come
out at its mouth, in which case, say the Namaquas, the lion invariably refuses to taste his
prey.

In further proof of this, Henrick the hunter
took up the tale, and said, "one night I was

asleep in my hut, when I was awoke by a noise outside. My wife whispered, 'I don't think that is a wolf;' on which I got up, and went out with a keree (or knobbed stick) in my hand, for I had no gun at that time. Below a tree I saw a cow lying, and as I went to it a large animal left the cow and came towards me. I stood my ground, and called out, when a lion, (which the large animal was), went off to one side. I went up to the cow, and found it, and another beyond it dead. The first had been ripped up, and the calf only eaten, because the contents of the stomach of the cow had come out of its mouth and nose, whilst the other cow had its neck twisted round, and its horns fixed in the ground, so that its mouth was kept in the air, to prevent the same ejection of food as in the first cow. I turned its mouth downwards to disgust the lion, and then went to sleep again."

In Henrick's word, during some months' acquaintance, I had implicit confidence; and it is quite possible that the lion may feel a peculiar disgust when the above accident happens, and to prevent it, could easily with his mighty paws, fix the horns of cattle in the ground. All this, if true, is a new and interesting fact in the natural history of the lion.

After my usual morning's walk of a couple of hours, I mounted, and in company with Kiusip on his large-horned ox, (and some of the horns of the South African oxen are six or seven feet between the tips), we were riding out of sight of the waggon, with a Herculean gun carrier on foot beside Kiusip, and looking among the bushes for game, when the chief, after casting his eyes to the ground, and taking his gun from his henchman, set off at a gallop, and I followed him rapidly, ready for a shot at game which I thought he had discovered. We continued thus for a couple of miles, brushing past the bushes, beside which I saw marks of human feet, when suddenly plunging into a hollow, we found ourselves close to three or four huts of stakes and bushes, and beside one of them was an old man, who seemed about to fly from us as his people had just done.

This was a Boschman, and Kiusip calling to him that there was no danger, he sat down on

the ground.   He was a wrinkled old man of five
feet six in height, and as wiry as an Arab; he
was darker than my escort, wore in front a flap
of soft black leather, and a little kaross of
springbok's skin was on his shoulders; his bow,
poisoned arrows, and assegay lay against one of
the huts; two little naked children crawled
on the ground beside him, and a third was
attending to a small conical shaped earthen pot
which, full of some green leaves, was cooking
on a fire.   This was the first Boschman family
we had seen on the journey, though it was only
part of one, the rest having nimbly flew at the
instant of our arrival at the kraal.

As we were to outspan about a mile only
from the Boschman, at a pool, I gave him some
tobacco, and invited him to come to me for
more.   And as we did not take the arms of this
first Boschman or murder him wantonly, to
which these people are sometimes subjected,
the news, as we afterwards heard, of our *cle-
mency* went continually before us.

The old man came to us in the afternoon.   I
gave him some flesh and a pipe, and throwing

the meat on the hot ashes, he watched its brown-
ing with impatience, and with restless eyes
glancing from side to side, then taking it up
when half dressed, he gave it a shake to free it
from the larger ashes, and putting an angle of
the meat between his sharp teeth, he sawed off,
with the blade of his assegay, a mouthful close to
his lips, (if his nose had been long the point of it
might have suffered), and so continued till he
had finished his portion, and then he contentedly
smoked, without apparently a care in the world.

> " Content doth seat itself in lowest dales,
>   Out of the dint of wind and stormy showers,
> There sit and sing melodious nightingales ;
>   There run fresh cooling streams, there spring sweet
>   flowers ;
>   There heat and cold are fenced by shady bowers ;
> There hath he wealth at will ; but this we know
> The grass is short that on the hill doth grow."

I asked him to take his bow and show us how
he crept after game; he, accordingly, with his
small weapon bent, and holding the slight arrow
in the middle between the first and middle fingers
of his left hand, and the notched end fitting on
the string between the thumb and closed fore

finger, and thus in a different manner from that of our Queen's Body Guard, the Royal Archers of Scotland, and other British Toxopholites, he, *in buff*, and quivering his arrow with his pointed fore finger, crept in a sitting posture among the bushes which overtopped him, and pretending to watch his opportunity, he shot his arrows at the supposed game.

He said that the poison of euphorbia, or milk bush (boiled till it was black), which he used, took from sun-rise to mid-day to kill the game (or about seven hours), and that after wounding bucks or other large animals, he leisurely followed on their spoor till he found them dead.

" When you are out hunting," I said, " and come unexpectedly on a lion, what do you do?"

" It is of no use to run away," answered the Boschman, " the lion would soon catch me, if he is awake and sees me."

" He is commonly asleep then in the daytime," I continued, " if he has nothing to eat?"

" Yes," was the reply.

" You sometimes share what the lion has got, if he is eating when you see him ?"

" Yes."

" Show me how you get part of the lion's food."

On which the old man, taking up his assegay, and walking backwards and forwards in front of a bush where a lion was supposed to be devouring a zebra or buck, and brandishing, but never throwing his lance, he addressed the lion thus, whilst he continued his to and fro walk—" What have you come here for? Have you got any thing to eat? You made such a noise I thought you had got something. Don't think to come here and quarrel with me, but go off now and get flesh." Thus walking and talking for sometime, he at last sits down facing the lion, when the astonished animal probably moves off, and leaves the remainder of his prey to the Boschman.

Lastly, I asked him to show me how he made fire; when he went and got two small twigs, and with his assegay he squared one twig and made a small hole in it, and gave the other a point, then taking out the bone barb of one of his arrows, he supplied its place in the reed with the pointed piece of stick, and placing the squared twig between his soles, he commenced rubbing the arrow-shaft between his palms, and pressing the pointed twig against the squared one, repeating the anxious cry of " hei! hei!" till he got a black dust from the two sticks, and after a quarter of an hour, smoke and a light.

At the conclusion of his hard work, I drew a lucifer match through sand-paper, produced an instantaneous light, and the Boschman was mute with amazement.

Continuing our progress over a level plain, on looking back, at a halt before sun-down, I was struck with the picturesque appearance of the party, and of the whole scene. A line of trees and bushes were on the right of the picture, indicating the chain of pools called Aansabib; in the foreground, on the left, were my own people and Namaquas on oxen and on foot; in the middle ground was a line of packed and loose bullocks with their drivers; behind them the sheep and goats; and in the distance the waggon;—whilst on the left, the view was bounded by tabular-shaped mountains, receding from us in broad steps; and overhead was a lurid sky

with heavy clouds slightly gilded with the setting
sun.

After a day's journey of six hours and a half,
we slept at two of the pools, and were for the
first time during the journey annoyed with brown
legged mosquitoes.   I, who had been for several
years in other lands subjected to the vexation of
these tormentors, could well appreciate our good
fortune, in having been hitherto free from them
on this expedition.

Beyond the Tarup hills, or " those containing
water in holes of the rock," and at no great
distance from us, was the fountain called Usis.

On the 22nd of March, we journeyed along
the chain of pools, saw the spoor of a cameleopard
with that of a Boschman after it, the fresh traces
of a rhinoceros, and not less than a mile and a
half springboks on the plain.   Some of the
hunters discovered a party of Boschmans on the
move, and they brought them (two men, six
women, and six children) to the outspan.

These Boschmans were well grown, and all in
good case, unlike in size or lankness the diminutive and starved creatures which are found on

the upper parts of the Orange river. The men wore karosses, and the heads of the women were ornamented with circular cut pieces of ostrich shell strung on the hair, one or two also wore dangling ornaments of red seeds. Their skin half petticoats were scantier, and their fore fringes shorter than those of the Namaqua women; otherwise there was a general resemblance in dress and feature.

The two old women of the group were very wrinkled and haggard, and had a look of the most perfect indifference to the world and every thing in it; the two middle-aged women sat with the children clinging to them, and seemed to be occupied with cares for them, and with the pipe; whilst the two youngest of the women, apparently about eighteen years of age, were tolerably good looking, and somewhat coquettish in their manner; but the whole would have been much improved by a plentiful supply of soap and water. Perhaps it is only during a shower of rain that water touches the bodies of most Boschmans!

White men they now saw for the first time, but they expressed no astonishment at our appearance. I showed them a looking-glass the first they had seen, and they seemed pleased with their own looks; and one of the men, about forty years of age, now discovered for the first time that he had a beard on his chin, and with which he seemed proud. I asked him what was that of all other things he wished for in the world—was it plenty of wives, of children, of cattle, of sheep, of clothes, or a good hut? and he answered, " the rhinoceros, and to get it easily."

I asked him, as I had asked the old Boschman (Ariseep) we had just parted from, to show me how he frightened the lion from its prey; and instead of walking about like the old man, this hunter started up, javelin in hand, and sprang about, fifty yards in front of a bush, with great animation, shaking his weapon and crying to the supposed lion, " What have you got there, cannot you spare me some of it? Be off, and let some stand for me, or I'll do you an injury," and then threw his assegay, but only half way to the bush.

Some Boschmans derive their principal sustenance from the lions: thus, Jan Buys said he was once in the field, and he saw a Boschman who answered Jan's inquiry as to his manner of living by saying, " I live by the lions."

" Well," said Jan, " there is the spoor of three before you."

" That is what I am now upon," said the child of the desert. " I let the lions follow the game, kill it, and eat a bellyfull; I then go near, throw about my arms and my skins, the lions go away grumbling, and I get what they leave. I never kill lions."

But Jan Buys heard afterwards that this same Boschman had been killed by a lioness: she was making a meal off a wild horse, and he did not observe she had whelps with her; beginning to halloo in his usual way, she looked up, growled savagely, and before he had time to retreat, she sprung at him, and destroyed him on the spot.

We moved on, followed by the Boschmans at a distance. It rained at sunset, and after dark we found ourselves under the trees of the Kei 'us, or Great Fountain. This rises in a broad patch of high reeds, at the commence-

ment of the Kei Kaap, or Great Flat; is a favourite resort for rhinoceroses, and large game
in general; and about it were old pitfalls for
securing them.

The Boschmans took possession of some bushes
about two hundred yards from where my kaross
was laid. I went to see them before I lay down,
and remarking a little child pounding a bit of
gum between two stones, apparently its only
food, I ordered the inside of a sheep (all I could
spare of it) to be given them, on which these
poor people made a capital meal. Nothing in
the shape of food comes amiss to a Boschman,
and every thing with blood in it (as a snake, for
instance) is eaten; and with them roasted intes-
tines are great delicacies!

In the middle of the night, when I usually
awoke to look round and listen, I thought I
heard a faint scream, but as it was not repeated,
I went to sleep again. On asking some of my
people in the morning if they had heard any
noise in the night, they said no, but that they
had now heard that the Boschman women
had been ill-used by our young Namaquas. I

went immediately to the head men, and talked very angrily with them about this disgraceful proceeding; but they did not seem to think it was a matter of any great consequence; and, as I saw one of the Boschmans quietly engaged looking about our fires for any scraps he could pick up, the only thing I could do was to give him a little tobacco and to follow the party, inwardly determining to prevent, if I could, a repetition of (what I then of course conceived) a very heavy offence, among any other Boschman families we might fall in with.

We naturally found the Kei Kaap or Great Flat, like all other great flats, exceedingly dull and tiresome; its surface was covered with thorn bushes, and here and there we passed rainwater in clay holes; between the Bulb mountains and a ridge west of us the breadth of the plain may have been twenty miles; on the right there was a notch in the range called " Isa koodee taos" (pretty girls pass.) In the evening we were rewarded for our dull day's journey by finding ourselves in the midst of scattered trees like those in an English park, with broad pastures and plenty of water: this was at the Huntop or Springing river, one of the finest in this part of Boschman land.

My four European followers cooked for me and for themselves day about, and the cook here incautiously filling his pot in the dark, we got in the soup substances which, between the teeth, felt like badly cooked liver, but which, on being held up to the light of the fire, turned out to be frogs!

As the place where we outspanned was said to be "sharp for lions," I lay down with my dust-man's bell at my head ready to sound an alarm: bells generally ring men to their dinner, but frighten lions from theirs. I neither lay in the waggon myself, nor allowed any one else to do so; nor had I the tent pitched whilst we were on the move, for as not a fourth part of the people could have been accommodated under cover, (by which they would have been safer from the attacks of Boschmans and lions,) I rather chose that all should be equally exposed, to prevent any grumbling or apparent favouritism.

It will naturally be supposed that I kept a

nightly watch, and I thought before I crossed the Orange river that it would be indispensible to do so; but I found that those keeping watch, and standing or sitting by the fire, would be sure marks for lurking Boschmans, and that lions pull out those soonest who may be sleeping with one knee up, I made all lie flat on the ground; first having made four half-circles of bushes to shelter us (divided into four squads) from the night wind, with fires at our feet, beside which the dogs lay to give warning of danger; the bullocks were driven up the last thing before we lay down at the fires, the better to protect them, and to allow them the honour of first falling in the lion's way. Of course we always slept in our clothes and shoes, with arms in our hands ready to spring to our feet on the smallest alarm.

On the 24th we were still crossing "the very desart." I saw no game, for it was all close to the foot of the mountains from which we were retiring. We reached the Arigha 'Oup (or Flowering Fish) river, which we crossed twice; and then on Sunday the 26th, we halted for some hours at the Oosep (or Foot) river, where we had a tremendous storm of thunder, lightning, and rain. The lightning played about the waggon in such a threatening manner (like fiery serpents) that I struck the boarding pikes, and laid the muskets flat on the ground, and for half an hour, the thunder crashed and rattled close over our heads, during the time I was endeavouring to lecture on the seventh chapter of St. Matthew.

Next day, we halted for the night at the Narop river, or that whose bed is free of stones, where, some time before, while a party of Namaquas were packing off in the evening, two Boschmans came to the opposite side of the river, shot their arrows wantonly at the party, killed a man, and made their escape. Perhaps the Boschmans had some former injury to revenge, and I thought that some of my party had lately brought on us all the risk of the poisoned arrow.

At the Narop the trees were covered with rats and mice busily engaged in eating the gum on the branches—some of the mice were striped.

On the morning of the 28th, several of the oxen could not be found, and it was supposed, for several hours, that the Boschmans had carried them off; but after a diligent search and a sharp pursuit, they were all recovered; and then zigzagging among low eminences, and losing our way under a stupid guide, a Bastaard dependant of the Buys, called "Oud Aaron," (who was good for nothing but watching a pot on the fire) we progressed over the endless Kei Kaap.

During the pursuit after the oxen I ascended a rise, and on looking towards the east, a reddish cloud extended wide in the horizon; calling on a couple of men to follow me, I was making for this, thinking it must be the Boschmans cooking one of the oxen, when one of the Namaquas said, "that is a cloud of locusts we see." Accordingly, turning again to the north, zebras were seen, and eight cameleopards; which last soon discovering the party, went off sawing the air with their long necks, like ships careering over the billows. As we were now on the tiptoe of expectation for the sport we should immediately find at the famed Bull Mouth's Pass, we did not follow on the spoor of the giraffes.

Sometimes outspanning without grass, and

sometimes without water, though seeing, on one or two occasions, appearances as if " the parched ground had become a pool, and the thirsty land springs of water;" but the deceitful prospect was occasioned by *mirage*, which, like a white cloud, reflecting the bushes upon it, lay spread on the ground. At last, on the morning of the 30th March, we saw a fine plain before us, which had been lately entirely covered with high grass, but which now exhibited broad bare patches; on looking to the right we saw the cause of this, for a red cloud, as of sand rising and falling, again indicated a thick flight of destroying locusts.

On the left of the plain was a broad and winding belt of high trees and bushes, indicating the course of a river, the Chuntop (or that which in running is suddenly checked): this entered a craggy opening in a flat range of mountains stretching across the plain to the north. The notch in the range where the wooded Chuntop disappeared, was the anxiously looked for Kopumnaas, or Bull's Mouth Pass—so named from its being full of dangers, like the Valley of the Shadow of Death.

I now girded up my loins for the chase, and I burned with desire to slaughter some of the larger game, as much to feed my fifty followers, who ate at the rate of two sheep a-day, as for mere sport. The people were divided into several parties, and we rode towards the foot of the mountains, where wild animals are always rifest.

We were not long before we saw a cloud of dust, which proceeded from a large troop of wild horses; dismounting, and extending ourselves, we approached them under cover of the bushes— they took the alarm—started off—passed through between us—galloped backwards and forwards— halted and gazed, and three fell under our fire in the course of as many hours hard exercise on foot. The moment the first, a full grown stallion, fell, and had stretched his powerful limbs on the plain, with the agonies of death in his eyes, half a dozen of the hunters collected round him; some of them brought dry sticks and made a fire; while the others cut him open, and taking the half-digested grass from his stomach, *they squeezed the moisture from it into their mouths* in the intensity of thirst; then cutting out the liver and roasting it, they made their breakfast off it; and, lastly, fitted themselves with shoes from the warm hide.

A troop of that most magnificent antelope, the koodoo, next occupied us for a little, but before we had time to secure any of them, we intercepted a dancing flock of springboks; and again, by sharp running and quick firing, three of them were also added to our larder. Our blood was now fairly up, and turning towards the mountain two large grey objects were seen, apparently disturbed by the " chattering of the musquets;" they ran a short distance among the bushes on the lower slopes, and then turned to look round them, these were two black and double horned rhinoceroses, covered with dried mud from the pools of the Chuntop, in which they had been wallowing.

We approached these dangerous animals with some caution, crept upon them, and got two or three flying shots at them; but unless they are taken standing, with deliberate aim at the backbone, or behind the jaw, good balls are thrown away upon them; not that their hide, though more than an inch thick, is impenetrable in other places to lead and pewter bullets, (hard and heavy), such as mine were, but because the rhinoceros runs away with a bushel of balls fired through his ribs.   In his side they seemed to make no more impression on him, at the time of receiving them, than so many peas would, though he may die from them afterwards.   So our two first rhinoceroses, being continually on the move, escaped from us though we tickled them roughly.

I now mounted my grey to look out for a good outspan place, whilst the locust-cloud passed over me, and the insects fell about me like thick and dry leaves in October.   I trotted leisurely among the bushes, admiring the picturesque entrance to the poort, with a noble pyramidical eminence, with three tiers of cliffs standing alone in the gorge, like a guardian of the pass; and which I named Mount Michell, after the Surveyor-General of the Cape of Good Hope; while dark precipices were on the right, and far in the pass was a great wall of rock stretching across it, towards which the trees of the Chuntop appeared to crowd, like Birnam wood marching to Dunsinane. I was enjoying this most romantic scene gleaming in the mid-day sun, when I remarked Kuisip, Henrick, and one or two more a little in advance of me, and looking earnestly towards the river.

I cleared the bushes, and saw rushing towards the hunters from the trees two more rhinoceroses, a female and a young male. The female appeared to have been wounded; for she snorted furiously, and driving her horns under a bush, she tore it up and threw it in the air, covering herself with dust and gravel, and then came on closely followed by her offspring, occasionally ploughing up the ground before her, and bent on destruction. The hunters now separated, and ran off as fast as they could to shelter themselves behind the rocks and bushes, whilst the monsters bore right down upon me, scattering a detachment of the pack-oxen as they neared me. Fortunately old " Night" was not paralyzed with terror, as some horses would have been, (and if he had stood still in the open ground on this occasion, we must have both been annihilated).

Wheeling him therefore to the right, I doubled the rhinoceroses, which with their deep-seated eyes and limited field of view, cannot see except right before them, and pulling up, I gave the dam a ball behind as she passed, which made her drop her tail, and the two then, tearing their way through a large bush, disappeared—
" Secouant la terre sous eux."

END OF VOL. I.

PRINTED BY WILLIAM WILCOCKSON, ROLLS BUILDINGS, FETTER LANE.

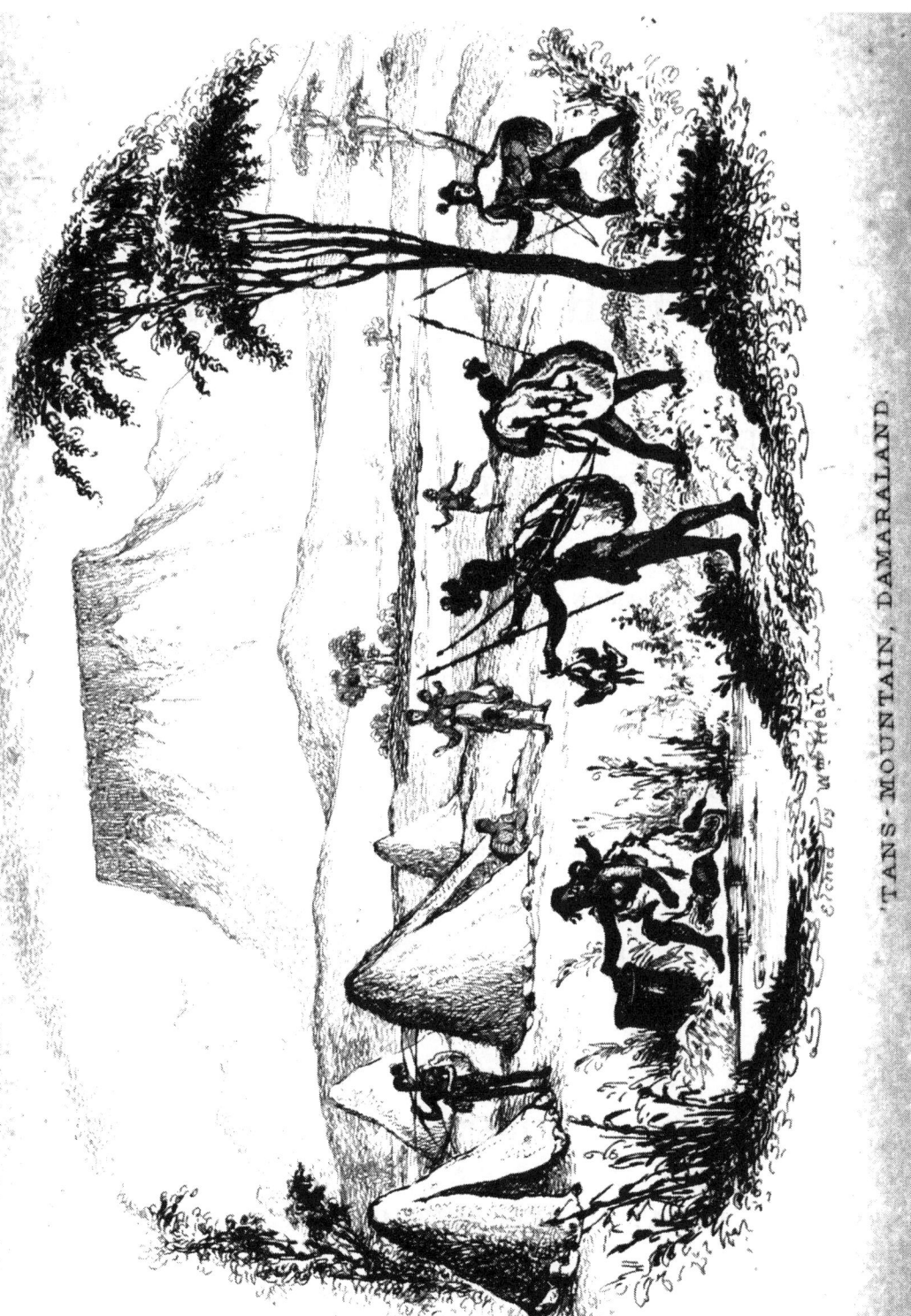

TANS-MOUNTAIN, DAMARALAND.

Published by Henry Colburn,13 Great Marlborough Street, 1835.

AN
EXPEDITION
OF
DISCOVERY
INTO THE
INTERIOR OF AFRICA,

THROUGH THE HITHERTO UNDESCRIBED

COUNTRIES OF THE GREAT NAMAQUAS, BOSCHMANS,
AND HILL DAMARAS.

PERFORMED

Under the auspices of Her Majesty's Government,

AND

THE ROYAL GEOGRAPHICAL SOCIETY ;

AND CONDUCTED BY

SIR JAMES EDWARD ALEXANDER, K.L.S.

CAPTAIN IN THE BRITISH, LT.-COLONEL IN THE PORTUGUESE, SERVICE,
F.R.G.S. AND R.A.S., ETC.

IN TWO VOLUMES.

VOL. II.

LONDON:
HENRY COLBURN, PUBLISHER,
GREAT MARLBOROUGH STREET.

1838.

PRINTED BY WILLIAM WILCOOKSON, ROLLS BUILDINGS, FETTER LANE.

CONTENTS

OF

THE SECOND VOLUME.

CHAPTER I.

CHAPTER II.

CHAPTER III.

CHAPTER IV.

CHAPTER V.

CHAPTER VI.

CHAPTER VII.

CHAPTER VIII.

CHAPTER IX.

CHAPTER X.

EXPEDITION OF DISCOVERY,

ETC.

.

CHAPTER I.

THE black rhinoceros, whose domains we
seemed now to have invaded, resembles in general appearance an immense hog; twelve feet
and a half long, six feet and a half high, girth
eight feet and a half, and of the weight of half a

dozen bullocks; its body is smooth, and there is no hair seen except at the tips of the ears, and the extremity of the tail. The horns of concreted hair, the foremost curved like a sabre, and the second resembling a flattened cone, stand on the nose and above the eye; in the young animals the foremost horn is the longest, whilst in the old ones they are of equal length, namely, a foot and a half or more: though the older the rhinoceros the shorter are its horns, as they wear them by sharpening them against the trees, and by rooting up the ground with them when in a passion.

When the rhinoceros is quietly pursuing his way through his favourite glades of mimosa bushes, (which his hooked upper lip enables him readily to seize, and his powerful grinders to masticate), his horns fixed loosely on his skin, make a clapping noise by striking one against the other; but on the approach of danger, if his quick ear or keen scent make him aware of the vicinity of a hunter, the head is quickly raised, and the horns stand stiff and ready for combat on his terrible front.

The rhinoceros is often accompanied by a sentinel to give him warning, a beautiful green-backed and blue-winged bird, about the size of a jay, which sits on one of his horns. When he is standing at his ease among the thick bushes, or rubbing himself up against a dwarf tree, stout and strong like himself, the bird attends him that it may feed on the insects which either fly about him, or which are found in the wrinkles of his head and neck. The creeping hunter, stealthily approaching on the leeward side, carefully notes the motions of the sentinel-bird; for he may hear though he cannot see the rhinoceros behind the leafy screen. If the monster moves his head slightly and without alarm, the bird flies from his horns to his shoulder, remains there a short time, and then returns to its former strange perch; but if the bird, from its elevated position and better eyes, notes the approach of danger, and flies up in the air suddenly, then let the hunter beware; for the rhinoceros instantly rushes desperately and fearlessly to wherever he hears the branches crack.

Thick and clumsy though the legs of the

rhinoceros are, yet no man unless possessed of the powers of my chief huntsman, Henrick Buys, can hope to escape him by fleetness of foot on open ground; once he has a man fairly in his wicked eye, and there is no broken ground or bush for concealment, destruction is certain. The monster, snorting and uttering occasionally a short fiendish scream of rage, bears down in a cloud of dust, tearing up the ground with his curved plough-share, kicking out his hind legs in a paroxysm of passion, and thrusting his horns between the trembling legs of his flying victim, he hurls him into the air as if he were a rag, and the poor wretch falls many yards off. The brute now looks about for him, and if there is the least movement of life, he runs at him, rips him open, and tramples him to a mummy!

In general, the moment a hunter fires at a rhinoceros, or hurls a lance at him from behind a rock or tree, he runs off as fast as he can, and if his gun is heavy, he drops it the better to escape to a place of safety, and from whence he can watch the movements of the rhinoceros.

By Behemoth, " the chief of the ways of God,"

is meant the rhinoceros. "Surely the mountains bring him forth, where all the beasts of the field play; he lieth under the shady trees, in the covert of the reeds and fens. Behold he drinketh up a river and hasteth not; he trusteth that he can draw up Jordan into his mouth." But the rhinoceros "which eateth grass as an ox" is the white rhinoceros, (which we had yet to see), larger but not so dangerous as the black species with which we had now to do.

When the elephant and the rhinoceros come together and are mutually enraged, the rhinoceros avoiding the blow of the trunk and the thrust of the tusks, dashes at the elephant's belly and rips it up. The lion of course never thinks of attacking the rhinoceros; and the Boschmans say that although found in the same haunts, they give way to one another. I thought then that the rhinoceros had no superior; none that he need fear save all-destroying man; when the Buys said—" Once at the Great Fountain, where we had gone to hunt, we found a rhinoceros which had just been killed by a hyena.

The hyena is in general a cowardly animal, and is scared even by a cow when it threatens with its horns in defence of its calf; but when the hyena is very hungry it seems desperate, and will attack any thing. The one which had killed the rhinoceros had followed it for some time, (as we saw by the footmarks), and had bitten it behind with its terrible jaws, till the rhinoceros fell and painfully died."

Having thus sufficiently introduced the black rhinoceros to my sporting reader, let us repose for a little under the refreshing shade of the trees of the Chuntop, before we endeavour to fight our way through the Bull's Mouth Pass.

*　　*　　*　　*　　*　　*

Many of the people were employed during the remainder of the 30th of March in *vleking* or cutting the meat of the game we had killed into thin flaps or steaks, and hanging it on the bushes to dry; and as "the ox is not muzzled whilst treading out the corn," so these fleshers' jaws were as fully employed as their hands. The most useless as hunters had the best appetites;

and old Aaron, the (so called) chief guide and butcher, ate till he could hardly waddle from bush to bush to cover them with red meat.

Aaron had a ten-pound appetite, or one of two ostrich eggs-calibre, and one ostrich egg is equal to four-and-twenty hen's eggs! If the reader is sceptical about the capacity of stomach of the Namaquas, let him see what the Esquimaux can do in the eating way, in the very interesting narrative of my friend Captain Back, and of other heroes of the pole.

In the evening I went with "oud Jan," Magasee, and three Namaquas into the pass to reconnoitre, and to look out for the best means of dragging the waggon through it. The valley at first was very narrow and rugged, with loose stones and bushes. Pathways cleared through the stones by the feet of wild beasts, led along the course of the river; and here and there, close to these paths, were circular enclosures of loose stones, about three feet high only, behind which the Boschmans had been in the habit of concealing themselves to hurl their lances into the bodies of the rhinoceros and other animals

as they passed.   We started steenboks, and saw
zebras grazing on a slope before us on our right,
and passed over the fresh prints of lions.   The
valley then opened, and became romantic and
beautiful.   It was of an oval shape, and three or
four miles in extent before it again contracted,
and it was full of various species of acacia, standing singly and in groups, whilst the mountains of
indurated sandstone which hung over it presented bare cliffs among scattered foliage.

We passed every instant the favourite resting-
places of the rhinoceros, whose disposition is so
spiteful that it kicks to pieces even what it de-
posits; and it seems always to return to the same
place for the purpose.

We crossed the Chuntop, and looking to the
left, I saw, eating its evening meal off a thorn
bush, a rhinoceros within one hundred yards of
us.   I whispered to Jan Buys, and we made
ready; but the watchful monster did not charge
as we expected, being young, and made off before we had time to *becreep* it.   After an hour's
progress through this first Valley of the Pass, (the
entire length of the Pass is about forty miles,) we

returned towards the outspan place. Two of the men in lagging behind, were pursued close up to us by a couple of old baboons, which had descended from some caves in Mount Michell.

In my absence in the Pass, many of the hunters had resumed the sport on the plain, and two more rhinoceroses were mortally wounded. The people ate apparently ten pounds of flesh each in as many hours; talked of their day's adventures, and how this one had ran off behind rocks or bushes, or how another had got into a tree, for fear of the rhinoceros; and with smoking, and pounding the bones on flat stones to lick the marrow, by drawing the stone across their lips: they were awake all night, made a noise like that in a shoemaker's shop with hammering the bones, and effectually kept off the lions or other nocturnal prowlers.

> " While the grim satyr-faced baboon
> Sat gibbering to the rising moon,
> Or chid with hoarse and angry cry
> The hunters who beneath him lie."

As we had plenty of wild meat for some days, I did not occupy myself next day on the plain in assisting to destroy wild animals, which we could not have carried off with us. I was very anxious to get through the Pass, to imprint its wild vallies for the first time with a waggon spoor, and to reach the sea as soon as I could, where, at Walvisch Bay, I expected to communicate with a ship of war, which had been kindly promised by Admiral Sir Patrick Campbell, for the assistance of the expedition.

In the morning of the 31st (having no time to lose in mere sport, though the temptation was strong), I mustered forty of the people, and proceeded with them to the left of Mt. Michell, to clear a road for the waggon. We cut down trees, lopped off branches, and removed stones for several hours, till the sun drove us in a fever of heat back to our " lay place." In the afternoon we inspanned, crossed the Chuntop, dragged the waggon down the first valley, again crossed the river, ascended a rugged slope, crossed a neck, and descending again on the other side, found ourselves in another beautiful valley, but of many miles in extent, running east and west, two or three miles broad, and enclosed with lofty mountains. At sun down, whilst outspanning beside a pool of the Chuntop, there was an alarm of a rhinoceros near the waggon; a few hunters ran to where he was; hunting frocks, jackets, and shoes, were cast off, and leather trousers were rolled up to prevent noise; the rhinoceros was *becrept*, the hunters sat down behind the bushes, long guns were rested on them and presented at the monster, which, unconscious of danger, was quietly eating from a bush, and three balls through the backbone and jaw, stretched the rhinoceros kicking in the dust—

" Procumbit humi bos."

In the night the dogs saved us from being run over by a rhinoceros, which was passing right through our lairs on its way to the water, but which, crashing through the bushes, was turned off by our watchful guardians. Next morning we proceeded with sharpened knives to cut up our mighty prize. The gastronomic powers of my people were so extraordinary that it seemed a rhinoceros only could satisfy them.

The huge grey and mud-covered mass of flesh

we now dissected, and whose hide we carefully removed and preserved, was a female, with two perfect horns of equal length, and she measured twelve and a half feet including the tail; inside we found a fœtus the size of a pig a month old. Aaron and his assistant butchers made slashing work, and we were soon in the midst of a great shamble; flesh, flesh was on every side, and the apparently insatiable stomachs of the Namaquas were at last content.

I went on in the afternoon with the head men, to reconnoitre our way in advance. We went west for some miles, and still following the course of the Chuntop, we turned north, and rode at a walk altogether eight miles. I found the passage for the waggon tolerably clear. It was quite beautiful—the great valley of the Chuntop: the mountains were between two and three thousand feet high, and of various colours and fantastic outline; smooth cliffs were on their faces, marked with white streaks, as if occasionally cataracts dashed over them. We saw some caves high up on some of the mountains, and the footmarks of a few Boschmans, but we saw no people. There

was an impressive air of solitude in this part of
the Pass, and an awful silence, interrupted occasionally by the strange cry, as of a lamb, of the
large blue and long-tailed colly.

We returned in safety, and passed the evening
beside the people.

" Tell me a rhinoceros story," said I to oud
Jan, the best story-teller of the party, and handing him at the same time a well filled stone pipe,
" we have had enough of lions for a time, now
let us discourse about even a greater than a lion,
the black rhinoceros."

Jan, as was the custom in the land, gave the
pipe to Saul, the Little Damara, to light (by
which practice boys early acquire the bad habit
of smoking), and then, after a few satisfactory
whiffs, he commenced.

" Once on a time my father took his sons out
to hunt; he only had a gun, and we had assegaes
and knives. At first we were very unsuccessful;
we found nothing till the second day; we were
very hungry, when we came on a rhinoceros.
The old man soon wounded it in the leg, and
he then told us to throw stones at it, to make the

wound worse. You know how Namaquas can throw stones; so we crept upon the rhinoceros, followed it, and threw stones with such effect, that at last it lay down from pain. I being armed with a knife, then approached it from behind, and commenced to hamstring it, while my elder brother, who is now dead, Cobus, remarkable for two strange rings round his eyes, tried to climb over the back of the rhinoceros to thrust his lance into its shoulder (it would have been very dangerous to have gone up to its shoulder on foot); he had just begun to climb, when the rhinoceros rose suddenly with a terrible blast or snort, and we all ran off as fast as we could to a tree, and there held a consultation about our further proceedings.

" We had not been long at the tree, when the rhinoceros observing where we were, rushed towards us with his horns at first in the air, and then as he came near, he tore up the ground with them. We scattered ourselves before him, when Cobus getting in a passion, stopped short in his flight, called the rhinoceros an ugly name, and turned and faced it. The rhinoceros, astonished at this unexpected manœuvre, also stopped and stared at Cobus, who then commenced calling out loudly and abusing the monster; it now seemed to be seized with fear, for it sidled off, when Cobus, who had a heart like a lion's and was as active as an ape, immediately pursued the rhinoceros, seized the tail, sprung with its assistance on its back, rode it well, and plunging his assegae deep into its shoulder, it fell, and was despatched by the rest of us. Hungry men can do extraordinary things—and this is a true story."

" I do not doubt it in the least," I said, " for I know that all the Buys have first-rate courage."

We retired to rest; but were roused in the middle of the night by a savage buffalo, which came from the eastern or upper part of the Valley, and attempted to run through us whilst asleep on the ground; but the ever watchful and faithful dogs turned it off also, and it was no more seen.

I had already picked up in the Pass the head and horns of a buffalo, which last, curved in a semicircle over the eyes, are very remarkable for meeting at the roots, and lying like a mass of rock over the forehead of this fierce and malevolent animal. Of greater size and strength than an ox, its body thinly covered with black hair, with a bristly beard about its mouth, its withers rising in a ridge, and with a short tail, the South African buffalo lurks in the thickets, and rushes out without previous warning on the passing hunter, and gores him to death. Some of the people had soles made of its enduring hide. I knew a man in Caffer land, a Hottentot hunter, Barber by name, who has gone up alone to a wild buffalo in a bush, and killed it with his assegae.

On the 2nd of April, we continued our progress through the Pass, first travelled west and then north, crossed the Chuntop three times, and after four hours, ascended a hill to the right, to cut off an angle of the river. We then went over some high ground among the mountains, passed some remarkable trees, eight feet high only, but six and a half feet in girth; the bark smooth and silvery, and the leaves oval. We

descended again, after some very rough shakings
for the waggon, and outspanned, after two and a
half hours of the mountain road, on the banks
of the Chuntop, where I was agreeably surprised
by finding many lofty fig-trees, fifty or sixty feet
high, and covered with ripe fruit, growing along
the course of the stream.

The stems of these trees were thick, numerous,
rather tortuous, and covered with a pale shining
bark; the leaves were entire, like those of the
ficus religiosa, and unlike those of the fig-tree of
gardens, divided into three parts; the fruit was
of the size of a Smyrna fig, and was very palateable; though I warned the people against indulging their appetites, for I was afraid of their
eating unripe fruit, and thereby producing dysentery.

On the following day, Elliot shot a dog-faced
baboon, five feet high; and four hours more along
the river and over a plain, freed us entirely from
the Poort, which we were very glad of, for the
labour we had undergone in getting the waggon
through was very great, accompanied with the
constant fear of seeing it shivered to pieces on

the broken ground it traversed; for it was more unmanageable than before, being now loaded with a rhinoceros' hide.

In looking back towards the Pass of the Bull's Mouth from the north, its savage aspect was very striking. The deep red precipices on either side seemed to enclose mysterious recesses full of danger from lurking Boschmans, lions, rhinoceroses, and buffaloes, whilst the conical summits of the mountain range rose high in the air, their sides deeply wrinkled with the water-worn furrows of ages.

I named two of the mountain peaks after my valued friends Dr. John Murray, Deputy Inspector of Hospitals, and principal Medical Officer, Cape of Good Hope; and Mr. George Thompson, the author of Eight Years in South Africa.

We discovered the traces of Boschmans, and set out in pursuit to catch them, to be our guides in advance. After a short time we secured a young man and a young woman, both very handsome! The man wore round his neck, attached to leather strings, a couple of pieces of ivory like paper folders, intended for eating a new fruit called 'Naras, which we had not yet seen. On his right arm, above the wrist, were many rings made of the hide of rhinoceroses, lions, kudus, and other wild animals, and worn as trophies; amongst them were distributed some teeth of the hyena apparently; and over his little apron, in front, was a disc of stiff leather, about three inches in diameter, and edged with iron, which looked like a miniature shield. He was armed in the usual manner, with bows, arrows, and lance. The young woman had a dangling bunch of red seed at the back of her head; she wore also the ivory scoops, and some leather rings on the left arm, also the common skin petticoat and fringe.

The Circassians, to defend their right arm, and to make it heavier in striking with the sword, wear a steel armlet; the Boschman, to make his arm heavier for throwing the assegae, wears his trophy rings.

The Boschmans having been freed from alarm by the grand succedaneum, the pipe, led us towards the river, where, beside some noble fig trees, in a hole in the bank, sat about a dozen

more of the tribe, men, women, and children,
eating figs and roasted locusts, whilst the flesh
of a young giraffe hung on a bush opposite the
den.

The water was in a deep hole under a wall in
the bed of the river, and the place where we now
outspanned is called Ababies, or Calabash Kraal.
The Boschmans here saw three new things,
white men, horses, and a waggon. Of white
men, they thought that they were not particularly
handsome, but, I fear, " rather the contrary;"
that is, we were thought to have been flayed!
My horses, they imagined were a sort of ox without horns, and said they supposed they would eat
very well; the waggon they believed at first was
alive, and afterwards, that it was one of the
strange white things (ships) which had come out
of the sea, and was now travelling over the land.

The Boschmans in the neighbourhood of the
Great River even, used lately to be very much
afraid of waggons: thus, Mr. Schmelen's people
once caught a Boschman, and he told them that
the first time he and his people saw the missionary's waggon they ran away from it for a

whole night, thinking it was some terrible monster, and that they always jumped over its spoor, and would not touch the wheel tracks on any account.

On another occasion, Mr. Schmelen sent out an old waggon with a hunting party, when one of the fore wheels was broken, and the waggon remained standing in the field for two months, at the end of which time a Boschman came to Mr. Schmelen's place, and said that he had seen the missionary's *pack ox* standing in the field for a long time, with a broken leg; and that as he did not observe that it ate any grass, he was afraid that it would soon die of hunger if it was not taken away!

I distributed small presents among the Boschmans of Ababies, and they seemed to put perfect confidence in us, and promised to show us, for a few beads and sticks of tobacco, certain watering places among the hills, known only to themselves, and lying between us and the Kuisip (or Root) river, for we had yet nearly three days journey to the river, and our road lay over an arid desert of sand, without any watering place

with which Aaron, the chief guide, was acquainted.

This old animal, a tall and thin Bastaard,
with a little flat hat like a crow's nest, and a
long-backed leathern jacket, afforded a good deal
of amusement to the people, from his appearance
and habits. I said before he was a ten pound
man as to appetite, and he was dirty and useless as he was voracious; his face had not been
washed since we left Tuais; perhaps that the mud
on it might preserve his complexion. His leather
crackers hung like a bag between his legs, and
so as to impede him in getting on his ox; and
it was said, that if he ever went to the Cape, the
first smith who saw him would strip him to make
a pair of bellows of his trowsers. Once on his
ox, he never got off either to hunt, to walk, to pack
an ox, or to assist the waggon, but with a skin
bag of dried meat behind him, he lingered behind and ate ever and anon during the march,
and at the end of it he sat down by the flesh pot.

To prevent a repetition of the proceedings at
the Great Fountain which had annoyed me so
much, I told the men of the Boschmans to come

with their families, at night, and to sleep within
sight of me, and that I would prevent their
women being troubled by the Namaquas, as
they might be, if they remained under the bank
of the Chuntop. To my exceeding surprise, im-
bued as I was with notions of Oriental jealousy,
the Boschmans said, " Take the women; the
people may do with them as they please; what
else is the use of them?" Seeing the Boschmans' feeling on this point (beasts could not
have been worse) I now thought that the occur-
rences at the Great Fountain were not of so
serious or disgraceful a nature as I had at first
imagined they were.

Can any state of society be considered more
low and brutal, than that in which promiscuous
intercourse is viewed with the most perfect indifference, where it is not only practised but
spoken of without any shame or compunction!
Some rave about the glorious liberty of the
savage state, and about the innocence of the
children of nature, and say that it is chiefly by
the white men that they become corrupt. The
Boschmans of Ababies had never seen white

men before, they were far removed from the influence of Europeans; the only thing they had ever seen of them were their ships, from the hills skirting the ocean, and yet observe what innocent notions they had !

Surely, all the restraints of civilized life are to be preferred to the licentious and shameless habits of the savage, and to such a state of moral degradation as was found to exist among the Boschmans with whom we had now to deal. Much need have these degraded beings to be taught better things; and though there might have been some excuse for their bartering their property in their wives and daughters for food, when they were starving, yet our Ababies acquaintances had not even this excuse, for they were well supplied with food, and were all in good case.

The locust clouds were about us again, and my people got sacksful of locusts from the Boschmans. The insects are caught at night by making circles of fire, within which the locusts fall; their legs and wings are then pulled off, and they are roasted and ground up fine. In the evening the Namaquas played at Hous in the bed of the river, squabbled over the game as usual, and skins being spread, they ate roasted locusts by handsful. I tasted them also, and they were sharp and bitter to my palate; it appeared like eating snuff. They would doubtless support life *at a pinch*. In the song of the wild Boschman he says,

> " I plant no herbs nor pleasant fruits,
>   I toil not for my cheer,
> The desert yields me juicy roots,
>   And herds of bounding deer.

> " Yea, even the wasting locusts'-swarm,
>   Which mighty nations dread,
> To me nor terror brings nor harm,
>   I make of them my bread."

I asked the Boschmans to dance, at sundown, by the light of a large fire, and they readily consented. The women drew up in line opposite the fire, and began, in a low voice to sing *ei oh!* *ei oh!* Clapping the hands at *oh!* they sang louder, when two of the men, stamping the ground for some time with one foot, and then changing for the other, circled in front of the women, and sang *oh, wawaho*, as an accompaniment, and their hands supported the body, which was occasionally twisted about; they also pointed at one of the women with their jackal's tail *hand-kerchief*, when she came away from the rest, stamped like the men, clapped her hands, and seemed to try and escape from them, whilst they continued to stamp and follow her round the fire. Several of the young Namaquas, excited by the wild strains, joined in the dance, threw their bodies into all sorts of contortions, and a scene of dust and noise ensued.

Not seeing the long nose of my Portuguese youth, Antonio, which was usually poked into any place where there was play going on, I inquired if he had come back from shooting: nobody had seen him since three in the afternoon, when he had been noticed some distance down the river, with his gun, and alone. I had repeatedly cautioned my Europeans never to go any distance from the outspan place without one or two Namaquas with them, in order that they might easily find their way back again, and for their greater security in every respect. I was very desirous that no accident should happen to the poor fellows who had trusted themselves with me so far; and, besides, it was my duty to care for them, in sickness or in health, in every possible manner.

Now it appeared that Antonio had left the Namaquas with whom he had gone out after midday, and had evidently lost himself, or some worse accident had happened to him. The dance was therefore immediately broken up, and parties went along the river for some distance, shouting and discharging guns; several fires too were made ; but no tidings were heard of him all night. In the morning I mounted a horse and rode in the direction where he was last seen, extending the people across the country to " cut his spoor," and expecting to find his lifeless remains. But, at a distance of two miles from the waggon we saw him approaching us; and it turned out, that though we were on an open plain, with scattered trees only along the river, he had, as some other Europeans would also have done, noticed the outspan so little that he had gone past it the previous afternoon, and wandering about looking for the waggon down the river, it fell dark, and seeing two of our fires at a distance, which he mistook for the eyes of a lion glaring at him, he had got up into a fig tree, to a branch of which he tied himself with his sash, and thus remained all night; but he could get little sleep in consequence of the cold, and had not heard our guns or shouting.

Next, in looking for our Boschman guides, we found, to our vexation, that the whole party had fled the moment the firing began as signals for Antonio. They imagined that they were about to be killed (as we afterwards learned), and accordingly took to the mountains, men, women, and children, leaving their giraffe's flesh on the bush, their locust sacks, jackals' skins, sandals, wooden hand troughs for drinking out of—in short, every thing they had except what was on them, and their arms, in the extremity of their terror: and now we experienced the painful consequences of carelessness.

Etched by William Heath.

EXPEDITION IN DISTRESS

Published by Henry Colburn, Great Marlborough Street, 1826.

CHAPTER II.

An unpleasant Prospect—Our drinking Water—Leave Aba-
bies—A false Report—A dry Outspan—State of the
Horses—A Poison Sucker—Mount Tans—The very
Desert—People and Cattle attempt to reach the Water—
The Bird of the Desert—Melancholy Anticipations—Send
off the White Men—A Dream—Some Relief—News of
the People—Abandon the Waggon—Death of a Faithful
Servant—Descend into the Devil's Den—Adventure of the
White Men—The last Horse Dies—The River Kuisip—
A Consultation—Excellent conduct of the Headmen of
the Namaquas—Missionary Influence—A disagreeable
Discovery—Story of a Boschman and a Lion.

THE prospect before us was now a most unpleasant one, we were at the last watering place
for nearly sixty miles; the horses and oxen
were thin and weak with their long previous
journey, the weather was very hot, and a desert
of heavy sand lay between us and the Kuisip
river, and whether there was water in it or not
at the point where we should first see it, we
were not aware. Still we could not stop where

we were, or go back; we must make a desperate effort to reach the Kuisip or perish in the at-
·tempt.

"Vestigia nulla retrorsum."

There was no retreating. If the Boschmans had not deserted us, we should have crossed the barren waste comfortably enough, for both men and cattle might have drank half way from holes in a rock, but where to look for these now, we were quite at a loss.

On the 4th of April, I found the Namaquas washing clothes, and some their bodies, in the only drinking place we had at Ababies. I am not very nice, but this was *too much*. However, I was now compelled to keep my temper, and to content myself with reproving the people for their excessive thoughtlessness. The oxen came late at night from the upper part of the river. I asked the cattle guards if they had been at the water with the beasts, and they said they had; but next day, when it was too late, I found out that the cattle had not been taken to the water at all, and early on the morning of the 5th, we left with thirsty oxen.

Our course was north-west, whilst the Chuntop
left us, and inclined to the west to disappear in
the sand before it reached the ocean. The
waggon moved slowly along, but frequently
stuck fast, and it was most painful to be obliged
to use the whip to the unfortunate cattle. After
five hours we got to a dry and nameless river,
and most of the Namaquas dispersed to look for
water but found none—thermometer 90°.

A report was now brought me that some
Boschmans were seen crossing the plain before
us, I immediately got on a horse and rode as
fast as the feeble animal could carry me for two
miles (forgetting all about poisoned arrows) till
I ascertained that the supposed Boschmans were
Henrick Buys and his gun carrier on the spoor
of wild horses, and thus our hopes of finding
guides to the water were again baffled.

I sat down for sometime on a hill, and waited
till the waggon and pack-oxen came on, and
though I saw a large flock of springboks below
me, no one was after them; water, not flesh,
was our only desire; the train, with two or
three drivers, passed on. I saw a few long-

legged plovers moving about, and after five more hours, we halted at half-past eleven at night in a valley of grass between low hills. The poor oxen were so knocked up with the heavy sands they had passed, with the heat of the day, and with thirst, that they could not touch a blade of the dry pasture, amongst which flights of locusts lay nestled for the night, here and there, and chirruped like young sparrows. I distributed all the water I could spare from the waggon among the people; for with their usual improvidence, the Namaquas had converted the goats skins, I had given them to carry water for themselves, into clothes bags, and now, consequently, they were reduced to extremity. I lay down thinking that to-morrow night most of the party are to perish.

Long before dawn, I was awoke from an uneasy sleep by Aaron the guide calling out to the people, "*Keiree! Keiree!* Rise! Rise! the sands are heavy, and the Kuisip is far off." I never got up with more uncomfortable thoughts. I rode to prevent the thirst as much as possible, but my horses neither answered whip nor spur; they were quite dull and very much knocked up, particularly the grey, poor Old Night, for whom I as well as my people entertained a great regard. I first made his acquaintance during the Caffer war of 1835, and up to this day he was always lively and alert. But my horses had a bad groom in Magasee, also first seen in Caffer land. I thought, from his previous life among the Caffers, that he would have been my best hand on the road and in the bush, but he turned out the black sheep of the party, a lazy, worthless fellow, who neglected his horses, and whose only enjoyment was sitting at the fire talking with old Aaron, "par nobile fratrum," pretending to preach after the manner of a missionary; and most strange to say, collecting the tobacco oil (a deadly poison) from his own and other pipes that he might suck it. This powerful narcotic made him, perhaps, more indolent than he would otherwise have been.

We moved slowly on, that is, the headmen and my own people with the waggon and pack-oxen, for the rest set off in advance in quest of water, and disappeared. Some hills showed themselves on our left, and far in the distance on our right, twenty-five miles at least, rose a great tabular mountain called Tans, or the screen, for it shuts in all the lesser mountains and hills near it; on each side of Tans extended black mountain chains.

Grey sand and gravel were around us on every side, and single blades of grass waved with the hot wind on the bare and burning face of the desert. The silence was deep and profound, for not a bird or insect was to be seen or heard. The poor cattle halting every few minutes, were ready to drop with heat and thirst, and tried in vain to bellow. After accomplishing a distance of about twelve miles slowly and painfully, the sand deepened so much that we could get the waggon on no further, and therefore we out-spanned; and as the only chance of saving the horses and cattle, I sent the whole of them off under oud Jan, after the people who had gone towards the river, and remained myself by the waggon with Kuisip, Choubib, Henrick Buys, and my white men. We had with us a few quarts of water to support us in this .

" Region of drought, where no river glides,
    Nor rippling brook with osier sides,
    Where sedgy pool nor bubbling fount,
    Nor bee, nor cloud, nor misty mount
    Appears to soothe the aching eye:
    But barren earth and burning sky,
    And the blank horizon round and round
    Spread—void of living sight or sound."

In the afternoon, whilst sitting under the shade of the waggon, which appeared in the midst of the desert of Tans like a ship cast away far at sea on a reef, we were visited by a singular little bird the colour of sand, and about the size of a lark, which ran round us apparently examining the strange visitants of the waste. One's desire of collecting was suppressed by the sight of this kindly and curious visitor. In the evening we went to the nearest sand-hill to the west, and hunted about for roots of shrubs and grass to make a fire, and collected sufficient to roast small pieces of meat for our supper. Expecting the oxen to come to us on the morrow, and placing our entire trust in Providence, we lay down at night and slept without being disturbed, though afterwards we heard that Boschmans were about us during the hours of darkness.

The whole of the next day we looked in vain for the oxen, we saw no signs of them, nor of any of the people; in the evening a gemsbok was tracked and shot about a mile from the waggon. We were now impressed with the belief that no water had been found in the Kuisip, where the people first reached it under a black mount which we saw in the distance to the north, and which I named after the well-known secretary of the Admiralty, Sir John Barrow (the distinguished South African Traveller, and chief promoter of Polar Expeditions); and that they being unable to help us, had either gone far down the river towards the sea in search of pools—or that they had miserably perished.

No more water remained with us than barely sufficient to support the life of two men for one day, and, as I felt myself bound "to stay by the ship" to the last, I told the three Namaqua headmen, that I intended sending away my white men to-morrow morning, to give them a chance of saving their lives, and that I intended (if no help came) to remain by the waggon till the water was expended, and if one of the Namaquas was willing to stay with me, to assist me in following the spoor of game to lead us to the water, which we might find about Mount Tans, distant eight or nine hours from us, I should be very glad.

After a consultation, it was arranged that Choubib should stay with me, and Kuisip, Henrick Buys, and my four white followers, in the meantime should leave us to shift for themselves, and in the hope of meeting again; but whether that would ever take place was very problematical. I comforted the people the best way I could, by telling them it was not unlikely the men who had left us were resting the oxen for a day at the water, before they returned with them to our assistance at the waggon; that to-morrow, when Kuisip and the others should leave me to look for water, and should happen to fall in with natives, their guns would sufficiently protect them; and that they must not now give way to despair, but exert themselves to the last.

" Ne cede malis, sed contra."

Secretly, however, I thought that our situation was most desperate. I also asked myself, " what

great sin have I committed, that we are now left
to perish, and that through my endeavour to
perform an extensive journey on the African
continent, so many poor people have been led to
destruction—what have I done to occasion so
much misery?—little hope there is now of ever
seeing home and friends again—our fate is sealed
in the parched and trackless desert of Tans."

Having all eaten a little biscuit soaked in
water (about the last of both we had), we lay
down to repose for a few hours beside the waggon
wheels. "Another night of misery," said one of
the white men to his comrade; and so it was,
for great thirst and anxiety made it so. In an
uneasy sleep, I dreamt that I saw a person spill-
ing water on the ground, and that I ran up to
him and fought with him for doing so.

At one in the morning of the 7th April, I got
up and sent off the people, with Henrick Buys,
carrying only their duffle jackets and arms, and
accompanied by my poor dogs. Kuisip, the excellent chief, whom I always found very quiet
and obliging, declared that he would not leave
me, but would stay with Choubib and myself, till

we also should abandon the waggon on the following morning, if no help came to us.

We tried to sleep again, and at sunrise, we three deserted mortals, were awoke by the barking of a Namaqua dog, which had joined us, and on looking up we saw two black objects approaching; thinking they were Boschmans, we made ready to fire, if their appearance was suspicious, but they came straight to the waggon without hesitation, and turned out to be two Damara men (slaves under the Namaquas), who had been despatched by Jan Buys with a small supply of water for us in the stomach of a sheep, and in the pericardium of a rhinoceros. We partook of this thankfully and eagerly, and then asked the news.

They said that the distance to the water was great, and that they had set out yesterday morning at sunrise, and had only reached us now;— that there were several high sand-hills between us and the river, and that it was impossible the waggon could get over even the first of them. It also now appeared that many of the people had nearly perished on the 6th. A number of

them had fallen down one after another among
the sand-hills, and with their skins dry, their eyes
bloodshot, a contraction of the throat, and their
mouth covered with a crust, they lay helpless
and dying. Some cried like children for help;
some were nearly blind; and others, mad with
thirst and the heat of the sand, had asked their
companions to make a hole and bury them, for
that they were dying, and could not go any
further. That Jan Buys and the stronger of the
party had gone on to the Kuisip with the cattle,
which had then been three days without a drop
of water. That the people and cattle, when they
saw the water in the river below them, ran down
as if they had been crazed, and cutting their legs
on the rocks, they scrambled down a steep precipice to reach the bed of the river, and throwing
themselves into the water, they lay in it, and
drank till the water ran out of their mouths
again; and that after this excess, some of them
had been attacked with a sort of cholera. That
Jan had then returned with skins of water and
pack-oxen to carry off those who had sunk down
in a dying state among the sand-hills, and had

saved them all! This last intelligence was very gratifying.

The Damaras having rested, I despatched them to the dead gemsbok near us, to get the water from its stomach. About mid-day we descried a dark mass descending a distant sand hill, and as it came nearer we found it to be people and cattle; shortly after the worthy oud Jan reached us; but he and the rest were low and dull. Jan confirmed the report about the impracticability of getting the waggon to the Kuisip, on account of the heavy sand hills and precipices about the river; said that though all the people at the river were alive, in the meantime, some of them might yet die from what they had suffered from thirst and fatigue, and from the way they had drunk water; that some of the oxen had fallen down the crags at the Kuisip, and had broken their backs; that some of the dogs were dead, and some of the sheep lost; that my horse Night was dead, and that England was dying; and that Magasee their groom had lost his riding ox and all his clothes, and was also nearly dead. Finally, Jan said that he had not seen anything of his brother Henrick or of the white men!

Henrick, the stalwart driver, inspanned his oxen; the waggon was dragged towards the river for three hours, when it "brought up," in the evening, at the first sand-hill. It was now, I thought, time to leave the waggon to the tender mercies of the wild people, who could not be far from us: I accordingly stripped off my hunting frock and worked hard for an hour among our stores; and, with the assistance of the people, I packed the ammunition, clothes, bird skins, &c., on pack oxen, leaving the rhinoceros' hide and whatever else could be best spared.

I now felt as if I were abandoning the wreck of my vessel, to which, as in duty bound, I had stuck as long as there remained a chance of saving it; and I thought that I should soon see it, from a distance, in flames, for the sake of the iron work which was about it. I cast a last look at our craft, with its tent sail died deep red for concealment, which was no longer possible on the bare sands; and I then followed the people and pack oxen up the sand-hill. Another melancholy sight now presented itself: poor Night, with apparently a number of dark Boschmans about him; but they turned out to be immense black vultures, which had already committed sad havoc on the head and stomach of my old and faithful servant.

We passed over no less than seven sand-hills, which were very steep. On the north side, and on their summits, were tufts of stick grass. Half way we met some of the Namaquas with a supply of water in the stomach of another gemsbok, and of which we gladly drank. After seven miles ride in the dark, we found ourselves on the brink of a precipice, and we looked down into a black yawning gulf, at the bottom of which, and about six hundred feet below us, glimmered a fire. This was in the bed of the Kuisip. The Namaqua head men wished me to sleep where we were, but I was so anxious about the fate of the white men, and Henrick Buys, that I resolved to make my way to the fire below.

Accordingly, staff in hand, and guided by oud Jan, I scrambled and slid down by a narrow, broken, and dangerous path, fit only for goats or baboons, the precipitous descent to the Kuisip. Jan being stout, got some heavy falls. When half way down, the people hearing our voices, set fire to a dry tree, to light us on our perilous way, and they were then seen running about in the red glare like demons in the devil's den— the name we gave this hole.

Elliot and Magasee soon after joined me to assist me down, and at the bottom I was very glad to find Henrick Buys, and Taylor, Robert, and Antonio, all alive, but lying down and very much exhausted. They said their joints were stiff with their walk in the night through the sands; and that with drinking so much water, in which they could not help indulging, their stomachs were quite out of order, and that they could eat nothing. I asked them how they fared after they left the waggon; they said they had lost one another in the dark, had wandered about, and had laid down in the sand-hills till sunrise, and in the extremity of their thirst they had been forced to resort to the last means to try and alleviate it, but that this had increased it.

Most of the bullocks were recovering, but the horse England which was below, was standing under a rock with some untasted grass before him. I lay down near him, and in the morning, when I awoke, not seeing him, I went up the deep and fearful looking bed of the river, enclosed with frowning precipices, in search of him, but he was nowhere to be seen. On returning to my kaross I found him stretched out dead, within a few feet where I had lain.

The river had not ran for some time, but in its bed were long pools of water, separated by sand and gravel banks. There were two or three thorn trees near the water, and on the shelf on which we lay, under a black and lofty cliff. The bed of the river was here so deep and narrow, that the sun was up for a long time before its rays could reach our den. Looking up and down the river for the short distance we could see for the sudden turnings, it was enclosed by the same

steep crags, whilst black and bare above us rose
Mount Barrow.

I went up and down the precipice I had descended the previous evening, and found it to
be composed of mica slate, the glare from the
shining particles of which was very disagreeable: I
was very thankful I had escaped with a whole neck.
I collected the baggage left here and there at the
commencement of the descent; ate a part of a
broken-backed ox; and then set off to visit the
waggon for the last time, with some people to
bring away a few more stores.   In the evening
I returned to the fire.

A deep consultation now took place among the
head men of the Namaquas, about future arrangements.   They saw I was resolved to reach the sea
at any sacrifice or risk, and they were well aware
of the value of the abandoned waggon, when, at
last, Jan Buys, of his own accord, proposed that
he should endeavour to save the waggon, and the
property left in it, by going back with it towards
the Orange river; and that, after a time, if he
heard no more of me, he should hand it over

to Mr. Schmelen, at Komakas. I was greatly
obliged by this very kind proposal of Jan's, which
afforded the only prospect of my ever seeing the
waggon again, though, of course, there were a
great many chances against my living to recover
it in any way. I promised Jan and Henrick
Buys two new guns for large game (the most
acceptable present I could make them) for their
great assistance to me; and which guns I should
send them from the Cape, if I ever reached it
myself again; and I inquired what more they
desired—a few beads, shirts, and handkerchiefs,
was all they asked.

Without the help I had already received from
the Namaqua head men, and particularly from
Jan Buys, with his span of powerful oxen for the
waggon, I could not have reached, at least,
nearly so easily as I did, the point I had now
attained beyond the Tropic of Capricorn. I had
no claim on these men for help or assistance;
they were free and independent in their native
land, and owed no allegiance to any superior. I
had come amongst them " a stranger and a pilgrim," with a few attendants, and no display of

any force to intimidate them, or of wealth to tempt their cupidity, or to induce them to expect great rewards for any services they might render me, and yet, seeing that I placed entire confidence in them, notwithstanding the evil reports of the people of the chief Abram, and knowing what help I stood in need of, they generously assisted me to the utmost of their power—Choubib, by becoming my interpreter; Kuisip, by bringing an escort to defend me; Jan Buys, with the use of his best oxen; and Henrick, showing where and how game could be killed to support the expedition.

I must not here omit to give all due credit to the salutary influence of the Rev. Mr. Schmelen over these men's minds, for I believe all of them had lived with him some time or other, and had, doubtless, benefited by the instruction and tuition of that excellent missionary. To missionary influence, then, I may say, that I now (through the grace of Providence) owed my life and that of the people with me. May that blessed influence be more and more diffused by instruments such as Mr. Schmelen is, a man not

bigotted to sect or party, or desirous of power,
but one with zeal tempered by a thorough knowledge of man in his barbarous state, seeking not
his own honour, but the good of his benighted
brethren, and making always every allowance for
their frailties.

On the 10th of April Jan Buys left me with
my waggon oxen and his own, twenty-five in all,
and ten men, to endeavour to retrace his steps
through the Pass of the Bull's Mouth (which he
had seen for the first time with me) and to reach
his place near the Orange river with the waggon.
I was very sorry to lose his company; for though
uneducated he had very good sense, was very
ready to communicate, and was agreeable withal.
As I had no companion, and was obliged to
preserve a certain distance with my people, not
from pride, which would have been contemptible,
but for the sake of upholding discipline, I used
to be much amused in listening to the long
stories of Jan Buys.

But I was rather anxious about the safety of
Jan and his men on their return journey, owing
to a discovery which I now made. When the

Boschmans fled from Ababies leaving their little property behind them, I requested the headmen of the Namaquas, to caution their people against taking away or injuring a single article belonging to the Boschmans, and I thought that the jackals' skins, which are of some value in the land, for clothing, and the other things had been left untouched; when I now found, to my disagreeable surprise, my Namaqua escort busily engaged in making fur caps and other articles of dress from the Boschmans' skins, which they had concealed from me till we had left the desert of 'Tans between us and Ababies. Now, the Boschmans are sometimes in the habit of poisoning water for wild animals or for *men*, when they want to gratify their revenge, and I was afraid that the water at Ababies might be prepared for Jan, when the Boschmans, from their mountain fastnesses, should note the approach of the waggon.

In the sequel what befell Jan after he left me shall be disclosed.

As this chapter on the escape of the expedition from a painful death, has been, I fear, rather a tiresome one, let us now finish it with a Namaqua story.

A Boschman was, on one occasion, following a troop of zebras, and had just succeeded in wounding one with his arrows, when a lion sprang out from a thicket opposite, and showed every inclination to dispute the prize with him. The Boschman being near a convenient tree, threw down his arms, and climbed for safety to an upper branch. The lion allowing the wounded zebra to pass on, now turned his whole attention towards the Boschman, and walking round and round the tree, he ever and anon growled and looked up at the Boschman. At length the lion lay down at the foot of the tree and kept watch all night. Towards morning sleep overcame the hitherto wakeful Boschman, and he dreamt that he had fallen into the lion's mouth—starting from the effects of his dream, he lost his seat, and falling from the branch he alighted heavily on the lion, on which the monster, thus enexpectedly saluted, ran off with a loud roar; and the Boschman also taking to his heels in a different direction, returned in safety to "his anxious parents."

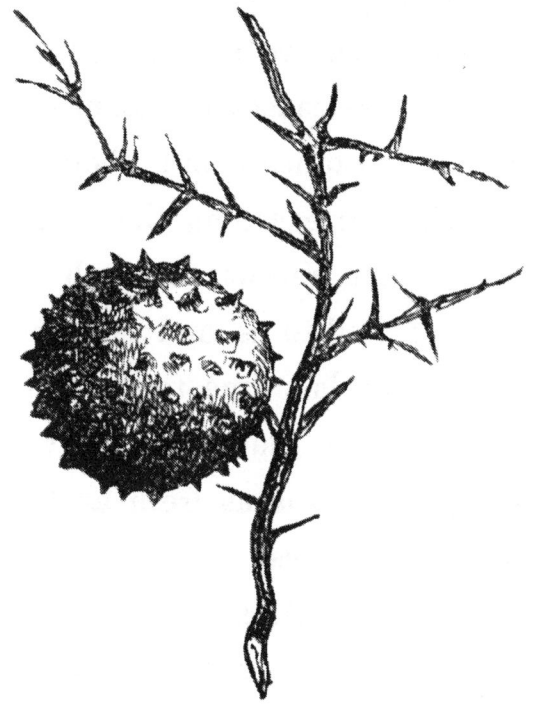

CHAPTER III.

WE had undergone heat, thirst, and much

hard work during the last few days, and it was to be supposed that we should now suffer from the effects of these causes of exhaustion, but in desperate circumstances, and when there is a great necessity for exertion, the mind sustains the body, so that unusual strength seemed to have been supplied to us, and I felt after I had descended for the last time to the Den to make the people bring up the cattle, that with a good bath in a pool, I was as fresh as I had been before leaving Ababies.

My favourite dog Moses (an odd name given him by the people) was not so easily restored to his accustomed condition, but stretching his handsome form, shaggy with black and white hair, by the side of the water, he lay incapable of doing more than looking up in my face piteously, whilst his eyes were glazing in death, occasioned from excessive drinking after exhaustion. I was forced to leave the poor animal to his fate with the carcase of England beside him, and cutting off the long tail of the horse as a remembrance of the Devil's Den, I scrambled up the precipice.

We had about fourteen oxen now to pack, and this work was therefore rather heavy; but by distributing people to the different packages, we got under weigh in about an hour.   There were only seven pack saddles from which leather cases hung, attached by straps to iron hooks; and the oxen which had no saddles, were packed in the Namaqua manner; that is, a boy held their heads by the thong of the nose stick; two or three sheep skins with the wool on them, were placed on their backs, by two men standing one on each side of the ox, a few turns of a riem or stout thong of raw hide (twenty-one yards long) were taken round the skins, and then against the sides of the ox were placed the packages, which were secured very tightly with the remainder of the riem, by the men placing their knees against the ox, and drawing the riem so tight that the poor oxen looked, after the packing was completed, as if they would be cut in two behind the fore legs; but "custom is second nature," and this tight lacing did not hurt them.

Our course was now along the Kuisip towards the sea, and on the evening of the 10th April, we packed off by moonlight, after eight miles journey over the sand hills of the south bank of the river.

Next day we accomplished the same distance only, as we were detained by a butchering operation of old Aaron. High cliffs still confined the bed of the river, and at a place where they were less precipitous, the oxen got down to drink. On the 12th, after twenty miles, we got a glimpse of heaven (as it were) in the river's bed below. Many acacias of pale foliage flung their arms over high grass of deep green, growing beside large pools of clear water; the path leading to this place of abundance was steep and rugged, but we managed to zig-zag down it, with the loss of another ox, which became incurably lame, from tearing off the claw or spurious hoof of one foot. It was killed, and the people having plenty of flesh for the nonce, made merry under the trees, eating, and drinking good water, and smoking. The oxen too once more got a good belly-full of capital grass.

The remains of a dead rhinoceros were found near us, which seemed to have been surprised

by the sudden rising of the river and drowned.
I am here reminded of a catastrophe which happened at a rhinoceros hunt, at which Henrick
Buys assisted, and, as we stretch ourselves comfortably by the fire in the evening, the men and
cattle lying refreshed beside us, after their late
struggles and privations, we may as well here
tell our story—

"Dulce est disserere in loco."

Henrick Buys was in the field hunting springboks, and having wounded one in the leg, he
followed it on the spoor with two or three other
men in company. They were coming up with
the game, when they crossed the fresh track of
a rhinoceros, and shortly afterwards saw a large
black male in a bush. Henrick immediately
" becrept" him, and with his long elephant rifle
he inflicted a severe wound on his fore leg. The
rhinoceros charged, the men fled, and the monster singling one of them out, closely pursued
him, when the man stopping short, whilst the
horn of the rhinoceros was ploughing up the
ground at his heels, and dexterously jumping to

one side, the rhinoceros missed him and passed in full career, and before the brute could recover himself and change his course, the whole of the party had got up into trees, whilst the limping rhinoceros was trying in vain to hunt them out by the smell.

The Bugbear in Jack and the Beanstalk, according to our Scotch edition of the story, says,

" Snouk but and snouk ben,
    I find the smell of earthly men;"

and so now seemed the limping rhinoceros to *snouk* or hunt about like a dog for his victims. One of the men, named Arasap, and armed with an assegae, said to his comrades, " Why are we all here doing nothing—shoot! shoot!"

" Well," said Henrick, " if you are in a hurry to shoot without waiting for the proper time, here is my powder-horn and ball-belt for you, and my gun is at the bottom of the tree."

Accordingly, Arasap descended from his tree, loaded the gun, and approaching the rhinoceros, he fired and wounded him severely but not mortally in the jaw; the ball was a leaden one, it did not break the bone, but was flattened against it, and stunned and dropped the animal.

The hunters now collected round the rhinoceros, thinking that it was incapable of rising again; and Arasap, in the pride of his heart, was directing the rest how to stab him with the best effect with their assegaes in different parts, when the beast beginning to recover, *spurtled* or kicked with his legs, and Henrick calling to the men to run for their lives, he set them the example, and swift-footed like Camilla, he scoured the plane, and was soon out of danger. The rhinoceros started up, singled out the unfortunate Arasap, and with ears erect, and screaming and snorting with rage, he thundered after him. Arasap, seeing that he was unable to outrun him, tried the same trick with which the other hunter had succeeded; that is, he stopped short, and hoped that the rhinoceros would pass him; the brute was not to be baulked a second time, but catching the doomed man on his horn under the left thigh (which was cut open as if an axe had been used), he tossed him a dozen yards into the air!

Arasap fell facing the rhinoceros, and with his

legs spread; the beast rushed at him, ripped up his abdomen to the ball-belt, and again threw him aloft. Henrick looked round, and saw Arasap like a jacket in the air. He fell heavily on the ground; the rhinoceros watched his fall, and running up to him, he trod upon him and pounded him to death. Arasap expired with the Namaqua exclamation of surprise and fear on his lip, " Eisey! eisey!"

After this tragedy, the rhinoceros limped off to the shelter of a bush. Henrick and the others crept up to destroy him. He dashed out again, and would have caught another man had it not been for a dog which came in the way barking. In turning short after the dog, the half broken bone of the rhinoceros snapped—it fell, unable to recover itself, and was immediately shot dead!

On the night of the 12th, we slept pleasantly, calculating on carrying the grass and water with us all the way to the sea, but on this journey, as on the great journey of life, if we were comforted at one time, we were tried with affliction at another.

Lest man should sink beneath the present pain,
Lest man should triumph in the present joy,
For, him the gracious laws of heaven ordain,
Hope in his ills, and to his bliss alloy.

On the 13th, our route lay in the bed of the
river over sand, and under the trees, which we
brushed as we passed along; we had much
trouble with the packs, owing to the oxen rubbing them off under the branches. The pools
of water appeared at longer intervals, and after
four miles they entirely ceased. In doubt and
uncertainty we accomplished twelve miles before
we came to where Kuisip was sitting by a small
hole under a rock, containing a very scanty
supply of greenish water, full of frogs and little
fish, and about which we saw the recent marks
of impure baboons; there was no water for the
cattle, and only a mouthful for the people.

Kuisip went in advance to look for water for
the morrow; we expected his return most anxiously, and looked for the sea from the cliffs,
but it was too far off yet to be seen. Kuisip came
back after dark, with the distressing news that
he could find no water whatever, after three

hours ride (twelve miles); we ate our supper in no enviable frame of mind, whilst the baboons howled in mockery, as it were, from the rocks above us. As we lay down under the trees, I thought "suppose to-morrow we go on for twenty-four miles (the utmost our oxen can accomplish,) and at the end of this distance we find no water, our strength will be exhausted, and we cannot return to this place, or retrace our steps further up the Kuisip; the best sight we could now see would be a pool of muddy water. I would give twenty guineas to any one who would assure us of finding water down the river, and within twenty miles of where we now are; to-morrow, may be our last evening!"

To sleep was impossible, for several hours; besides melancholy anticipations, the troublesome insect, the bush louse, with flat red body and streaked legs, attacked us without mercy, and the baboons *quahed* horribly, close beside us.

At daylight there was nothing for us but to pack up and to push on as vigorously as we could. Again the oxen annoyed us by running under the branches of the trees for coolness, and some lay down exhausted, and had to be unpacked before we could get them up again. The banks of the river became lower. On the right were rocks of mica slate, and on the left sand hills. The dubbee boom, or tamarisk tree, apparently the type of this part of Africa, and which I had constantly seen from the Kousie to the Kuisip, was now covered with white bloom. Of the chief Kuisip, who took upon himself the charge of searching for water, we heard nothing; where he was, and whether he had found any water, we, bringing up the rear of the people and baggage, knew not. After one-and-twenty miles, the oxen seemed quite "done up" with the heavy sand in the river's bed, the heat (93°), and thirst. They had no water the night before. I halted the distressed cavalcade beside some reeds, where the cracked clay indicated a recent pool; we dug here but found no moisture.

I was undecided whether or not to pack off and rest the oxen a little, though it appeared our case was nearly hopeless, when the chief's gun-carrier Einap (or liver) appeared with a smiling countenance, and pronouncing the magic word 'kams (water), the people set up a shout of joy, and most of the Namaquas leaving us to get on the cattle the best way we could, set off to refresh themselves. After considerable trouble with the bullocks, we got them on four miles more (twenty-five in all), when, below a sort of step in the river's bed, among a large patch of reeds, there was found, to our exceeding relief, a good supply of water.

This place was to us, parched and hard wrought as we were, a little Paradise. I felt again quite contented, seeing how the people and cattle were enjoying themselves, reposing under the trees and among the reeds. On the reeds, by bringing two or three together, the red-headed weaver bird had hung its light grassy nest, which waved in the air with the wind. In the evening, I went up the hills on the north bank, to look for the sea which we were striving so hard to reach; but I could see nothing of it. Bare and extensive plains lay to the north, and at my feet were large crystals of hornblende imbedded in quartzose rock.

We had not got any game for some days; the sheep were almost all eaten, and the broken-backed and lame bullocks devoured by my forty followers. Not knowing that we should obtain any supplies at the sea, (and we were almost certain we should find no game there), we were now reduced to very short commons. A sheep was made to go a long way, and none of us had ever sufficient to appease our hunger. The Namaquas asked for a bullock's hide, which we had kept to make shoes of, and roasting it at the fire, they pounded it between stones, and devoured the whole of it. I partook of it also, and found it very tough, but not disagreeable to the taste: to be sure, at the time, I could have eaten my saddle for hunger; and I certainly thought that our leather trousers must soon furnish a meal. Old Choubib was a great talker and a great eater; and when he got a mess of meat before him, he made always a large hole in it. An ingenious device was fallen upon to cheat him of his usual portion. When he sat down to eat, one of the white men asked him a question on some subject, he answered it at length; then another would ask his opinion on something else, and thus he would be kept talking whilst the rest were busy eating from the mess; and when he had finished his discourse, he found but a scanty morsel left.

We halted four-and-twenty hours, and then went on again. Shortly after we left the reeds, we saw the footmarks of men. Many of the Namaquas got alarmed at this, and wished to pack off that we might ascertain who and what the strangers were. We had many stories among the people of the wild men who lived by the sea at the mouth of the Kuisip, of their killing white sailors, of their bloody battles with the Damaras, &c. But I would not consent to delay: we went on, and after fifteen miles march, having lost all the cliffs and crags which higher up had enclosed the river, we offpacked on its grassy bed. By moonlight we saw a place which looked damp, and digging there we luckily found water.

On Sunday, the 16th, we were obliged to hurry on. Trees and grass were plentiful in the broad bed of the river, but no water was seen. Sand hills continued on our left, and increased in

height and in variety of outline.   On our right,
was a plain covered with granitic sand; and
bearing north from us lay a high mountain,
apparently ten miles off, and two or three thousand feet high, which I named Mount Hamilton,
after the worthy President of the Royal Geographical Society, W. R. Hamilton, Esq., F.R.S.

After sixteen miles journey, we halted at a
place where huts had lately been erected, and
where we got dirty water by digging for it.   In
the evening, a broken murmur, borne up the
river by the west wind, broke on our attentive
ears.   This was the roar of the breakers on
the coast, and though we were yet far from
the ocean, its music was most soothing and
delightful.

> "As some lone traveller, who the livelong day
>   Toils in the sandy waste, or fainting climbs
>   The lofty mountain, and in distance views
>   Gay smiling fields and turrets tipt with gold;
>   His ravish'd soul exults, refresh'd he breathes
>   The purer air of Heaven, and pursues,
>   With double vigour, his meand'ring way;—
>   Thus did our ear inhale the blissful sounds,
>   And our heart beat with a redoubled joy."

Next day I thought we were to lose Elliot. He had hitherto borne his fatigues well, and was indefatigable as a sportsman; but now, shortly after we started, he could neither walk nor ride his ox, feeling so sick and weak, and he lay down on the sand quite exhausted. With some trouble I got him on; but after eighteen miles we were all obliged to halt, the cattle could travel no further; and as we had not a drop of water, (our two little kegs, canteens, and stone bottles were empty), the people were plunged again into the depths of despair.

Leaving them to sleep away the sense of their present misery, I wandered about the broad bed of the river with Henrick Buys for an hour or two, looking earnestly into every patch of reeds or long grass for moisture, and digging with our hands in the clay and sand at the most likely places for finding the indispensible element; but no water could we find. With our mouths as dry as a dusty road, and hardly able to speak, we looked about for some green grass to chew; and, to our most agreeable surprise, we found the new fruit 'naras, of which I had first heard from the Boschmans of Ababies.

The 'naras was growing on little knolls of sand; the bushes were about four or five feet high, without leaves, and with apposite thorns on the light and dark green striped branches. The fruit has a coreaceous rind, rough with prickles, is twice the size of an orange, or fifteen or eighteen inches in circumference, and inside, it resembles a melon, as to seed and pulp. I seized a half ripe one, and sucked it eagerly for the moisture it contained; but it burned my tongue and palate exceedingly, which does not happen when this most valuable fruit is ripe; it has then a luscious sub-acid taste.

Kuisip had not yet appeared, and he was always most active in his endeavours to find water—till he returned with bad news I could not despair; and, though of late, our circumstances had often, apparently, been most desperate, yet

"Hope springs eternal in the human breast;"

and I had always vouchsafed to me an under

current of that consoler to afford some comfort, and to buoy me up.

I returned to the people, who had all thrown
themselves under the trees and bushes, and
were trying, by keeping quiet, to prevent their
thirst increasing to a dangerous pitch. The
goats were milked, and a few spoonfuls of milk
were distributed as far as it went, but this small
supply was of no use after a hot march under a
temperature, in the shade, of ninety degrees.
We also tried to eat a little, but it was impossible
to swallow the food.

When matters were in this miserable state,
I saw Kuisip approach, with two or three of his
people; I hesitated at first to ask him the news,
but at last said, " Is there water?"

" Yes," he answered, " and we passed it on
our way here."

" Is there enough for the men and cattle?"

" There is enough."

On hearing which the poor people's eyes,
which had been clouded with despair, immediately brightened, and they gave themselves up

to joy. " 'Kams; 'kams!"—water, water, was heard on every side.

To compare great things with small, as the soldiers of the gallant Moore, languid with their distressing retreat, rose fresh from the heights of Corunna with the prospect of a contest, so did my people, exhausted with their thirsty journey, acquire a new life with the prospect of moistening their parched bodies.

Little did our friends at home then suppose, that we were delighted beyond measure at finding two little holes full of muddy water. Ye, whose tongues have clove to the roof of the mouth with thirst, can appreciate the exceeding relief we now experienced at the immediate prospect of swetting our cracked lips!

On the 18th of April we were just three months from the warm-bath. I now thought that I could not wish my worst enemy more trials and troubles than we had experienced in that time. Still I was very thankful that myself and people were yet alive—that I was now quite free from lameness: and I hoped that the journey, long as it had been, would not prove to have been undertaken in vain—that our labours would meet with the approbation of our countrymen, and would be eventually attended with benefit to the human race. One thing was unpleasant, I could not help overhearing, among some of my white and coloured attendants, expressions indicative of their being tired of the journey, though as yet, they did not complain to me of it; but after what they had suffered, it is not to be wondered at, that they thought they had had enough of the south-west coast of Africa.

" Come, men," I said, " we must not lose heart—we have already got further to the north of the Cape than any other white men before us —we must persevere and try how much more we can do—we must not be laughed at on our return."

In the afternoon, we reached Aban'huas, or Red-bank, a part of the river so named from the red colour of the sand-hills on the south-side. Here we found a deserted hut, of a conical form, and composed of stakes and bushes, and beside

it, among reeds, there was excellent water.
We again saw the recent spoor of men, and in
order to obtain guides and supplies of food, it
became necessary to hunt up the people. Ac-·
cordingly, after a pursuit behind, and among
the sand-hills, two heads were at last seen peep-
ing over a knoll, and our Namaqua pack of
hunters, by circumvention, soon secured two
stout fellows.

Our captives belonged to a large tribe of red
men, speaking the Namaqua language, and who
inhabit the shores about Walvisch Bay. They
were tall and good-looking for Namaquas, and
wore fur-caps, handsome mantles of jackals'
skins, ivory scoops about the neck, trophy rings
of leather round the wrist, the disc or circle of
leather in front, and sandals on the feet. They
were quite ready for action with bows bent,
quivers of soft leather full of poisoned arrows,
and lances. And for provisions, they carried at
their backs nets containing half-a-dozen of the
ripe 'naras fruit, which served them for food and
water.

These two men were spies, who had been sent

to reconnoitre us from the main body. At first, they were in some trepidation, seeing the number of guns we had, but on being presented with a pipe and a piece of flesh, and being assured that they had nothing to fear from us, and that we merely wished to go to the sea to look for a ship I expected, and that I wished to purchase some provisions from them, and not to take from them their property, they became composed, and I asked them the news.

They said that it was now the commencement of the mist rains at Walvisch Bay, when the ships arrive to catch whales; that no ships had been there for a long time, but that they now expected them every day—that the Damara negroes of the plains were at the distance of a month from them, in the upper parts of the Swakop or Bowel river, which, like the Kuisip, emptied itself into Walvisch Bay—that they had no friendly intercourse with the Damaras, of whom they were much afraid, as they were a strong people, *and very angry*—that once they had gone up the Swakop, on a hunting expedition, and had got under a high rock, on the top of which were Damaras—that instead of the Damaras shewing any desire to be friendly, they shouted, and threw down stones at the Namaquas of the Bay—that beyond the Damara negroes, and along the coast, is another nation of *red* men, called Nubees, or the Many People, and which people are friendly to strangers—that it was impossible to get to them now, though the chief of the Bay had once visited them, but he was now absent on a visit in the interior, and no one else at present at the Bay could undertake to shew the waters beyond the Swakop.

" Besides," said the spies, " we are always afraid of meeting the Damaras on the sea-shore, to which they occasionally come on their hunting expeditions, after the *elephants* and other large animals in the Swakop. Not long ago the Damaras came down and attacked the people of the Bay, who at first fled; but watching the Damaras as they separated to eat the 'naras fruit along the Kuisip, we killed a number of them, and the heaps of stones you passed the day before, are their graves: after this the Damaras have not troubled us."

The spies had heard of our approach from a Boschman who had been near the waggon when it stood in the desert of 'Tans, and who had heard the shots fired when the gemsboks were killed. The Boschman came along the river, and told the people who were lying in it, that a large commando, or armed party, was coming against them to plunder them; and they accordingly left the river and fled among the sand hills; but the chief's wife, who was left at the bay, told her people not to be alarmed, or to run away, but to collect the cattle and sheep, and see what assistance could be given us.

I was much surprised and pleased on hearing the friendly intentions of the chief's wife; and I immediately dismissed one of the spies with a present of a large handkerchief for the head of the lady, and with tobacco for her pipe; and I directed the messenger to say that I hoped to meet with her at the sea in a day or two, and that she need be under no apprehension of any evil from us; for we were merely hunters of game, and not robbers of cattle.

In the afternoon we packed up and went

along the river for some distance, then left it to the right, and got amongst sand downs; and some time after sundown, we packed off for the night at two or three huts at a distance from water, but surrounded with heaps of 'naras skin. Here we saw a few new men's faces, but no women.

The huts were of singular construction. Crooked stakes were arranged in a circular form, and met at the top, where a stout straight post supported the roof. Some of the crooked stakes projected beyond the entrance, so as to form a porch, to prevent the west wind from blowing into the hut, which was well thatched with grass and reeds, and was roomy and comfortable inside.

To prevent the oxen straying among the sand hills, we attached them by the nose thongs to the packing riems, stretched between the cases; and scooping out for ourselves beds among the sand, we lay down to sleep in peace, seeing that we had secured the good will of the people of the Bay, who have got the character in Namaqua land of being a very wild tribe.

On the 19th of April, after allaying our hunger

and thirst with some ripe 'naras, the entire support of the Bay people for two or three moons or
months—at least, so they gave me to understand
—we continued our march among the sand hills,
and on descending a high one, a plain covered
with reeds and grass was spread before us, on
which were hummocks of sand covered with
bushes, and in the horizon gleamed the welcome
ocean, now reached for the first time at this
point from the Cape, from which it is distant
12° of latitude. We halted at a number of
empty huts, near a pool of brackish water, and
pitched our tent not far from Pelican Point,
Walvisch Bay, in lat. 22° 55′ south.

CHAPTER IV.

WALVISCH BAY is a considerable indentation
in the line of the west coast of South Africa, its
length from North to South, along the coast,
may be about twenty-five miles; the most secure
part of the bay is that behind Pelican Point, (a
long spit of sand, alive with wild fowl), which
prevents the west wind rolling the billows of the
Southern Atlantic over the anchorage behind

it. There is a broad sandy beach round the bay, and sand hills heaped up in various forms inland, and the general aspect of things here is very wild and Arabian-like.

Where we lay, was two or three miles from the south end of the Bay; and on the afternoon of our arrival we walked with eagerness and impatience to hail once more the ocean, which having now reached, we thought we should not perish. We expected to see ships, and to find stranded fish, or shell fish, to support us. Along the shore of the shallow southern creek were long lines of dead mullet and cat fish, and at two different places we saw the skeletons of human beings, half covered with sand.

We thought of the stories we had heard, of white men having been cut off at Walvisch Bay; but, on questioning our guide, one of the spies we had caught at Aban'huas, he said that one set of bones we saw belonged to a feeble woman, who in wading into the shallow water to fish, had stuck in the mud, and was drowned by the rising tide, and that the other bones belonged to a man of the bay, who was lamed from a fish-

bone running into his leg, and who fell and
died one day on the sand heap where we saw his
remains.

But afterwards we found out that these bones
were actually those of white men. A woman
told one of our Namaquas, that a captain of a
ship, who was called by her "Hous," in returning to his boat, was assegaed on the beach, his
men having interfered with some of the women;
and that from a similar cause, and on another
occasion, when a whale had been struck, and was
lying stranded near the mouth of the Swakop,
two boats' crews landed near it in the evening to
cut off the blubber, and that the bay men, with
broken assegaes concealed under their cloaks,
mixed themselves with the white men, and
watching their opportunity when the sailors were
sitting by their fire at night, they rose and
stabbed them all except one man, who escaped
up the river, but who was also killed a day or
two after.

It is very difficult to restrain "the wild spirits
of ocean," on first landing from a voyage; but
it ought to be the chief point of attention of

masters of vessels to prevent their men irritating the natives any where, by interfering with what does not belong to them, and not set their men a bad example as I have myself witnessed. Among barbarians a stranger ought to have the greatest command over himself, and be under strict self control, or there is a chance that he will not travel far, or long sojourn in safety among them. If his moral principle is not strong enough to control him, and to cause him to refrain from committing a great sin, then common prudence ought to dictate to him not to covet what is his neighbour's, when he places himself completely in that neighbour's power. Savages have affections and feelings like other men, and all are not like the Boschmans of Ababies; let the white stranger then ask himself, if tempted to try seduction by beads or toys, how he would relish that those he left at home should be tampered with, or be induced to violate their pledges made to *him* who now seeks to inflict a mortal injury on another. We see daily instances of retribution in this world, and is it not likely that in the case of a seducer especially, he

will one day writhe under a similar injury to
that which he may now recklessly inflict?

' Let this rule be recollected in travelling, " If
the jealousy of savages is roused, they immediately become most implacable enemies, and
even if they are condescending in a particular
way, that condescension being taken advantage
of, places them on a level with you, and destroys
your superiority over them." Many an expedition carefully prepared, and which may have
started with every prospect of success, has been
ruined from this cause alone, interference with
women, though this, the true cause of an expedition's failure, may not have been revealed to
the world.

The quantities of sea fowl we saw on the shores
of the bay, winging their way, and screaming over
its green waters, were immense; pelicans with
snow white plumage, and a slight blush of red on
the wings, appeared in vast flocks; flamingoes with
out-stretched necks and drooping bills, stalked
along the beach, and not having been fired at for a
long time, they allowed us to approach them; wild
geese in long strings flew over head, out of reach

of our guns, and sand larks, useless to us, owing to their diminutive size, hurried along the wet sand before us.

Substantial food was what we craved, and a dead fish we had no objection to, provided it was not too far gone. At last we got a great prize in a stranded cabaljao, fifty pounds weight, like a huge salmon, and which took two men to carry it on a stick between them; by the gills it appeared not to have been dead many hours, and had ventured too far into shoal water after the small fry.

Further on we fell in with large muscles of excellent quality, and digging with our hands in the sand, we collected a quantity of clams. This change of food was to myself and my white men a very great treat; all the biscuit was used, and of course we had had no vegetables, and it is only after much rain (which we had not yet experienced) that roots are to be found; of flesh we were quite tired, and though we had not got enough of that of late to keep us in proper condition, I myself was disgusted with its endless repetition, and yet I believe that our Namaquas

not caring for fish, and even disliking it, would
have willingly gorged themselves with flesh alone,
from one year's end to the other, if they could
have got it.

The bay people catch and eat fish after
the 'naras is out of season, and the carcases of
whales, killed by the crews of whaling ships,
afford them savoury repasts in the months of
May, June, July, and August, or during the
time the whalers are about the bay. After this
they hunt, obtain roots after rain, and kill an
occasional heifer or sheep, till the 'naras season
again comes round. Thus they make out the
year without cultivation of any sort, not even
melons or tobacco, of which last they are ex-
travagantly fond, two or three sticks being the
price of a sheep.

After walking about ten miles from the Tent
we were opposite Pelican Point, on which we saw
the jaws of whales set up like beacons. We
looked in vain for any post or staff erected to
tell us that a ship of war had visited the bay to
assist us, and would come again, but I hoped that
we should soon see one of Sir Patrick Campbell's

squadron, by which I might be set down at
Benguella, and from thence travel east. Since
my prospects of getting further to the northward of Walvisch Bay by land were at present
bad, I resolved to tarry some time at the
bay to give a chance to the man-of-war to
arrive, and if in ten days or a fortnight she
did not appear, then I intended to penetrate
to the eastward from the bay, as far as I
possibly could, hoping that I might not be
"brought up" till I found myself in the Mozambique Channel; remembering that we must
strive continually to.

> "Conquer difficulties
> By daring to attempt them—Sloth and folly
> Shiver and shrink at sight of toil and danger,
> And make the impossibility they fear."

I returned to the tent, round which the Namaquas were sheltered behind screens of bushes
and reeds, and seemed to be comfortable; and
I was glad to find some of the Bay people beginning to occupy the deserted huts; but as yet
none but very old women appeared, besides a
couple of dozen of stout fellows, (some of them

in penguin caps) who went about always armed
and prepared in case of treachery on our part,
though as we saw neither flocks nor herds, there
was no temptation to molest them. Whilst we
slept with arms in our hands and the dogs at our
feet as usual, in case of a night attack. The old
women, who wore the usual skin petticoat, a
flap behind and fringe before, tried to render
themselves attractive with cowery shells hanging
over their eyes, and with rosettes of the same
sewn on leather, and attached to one side of
their head, " we are willing to find husbands
among your people," said the old dames !

I thought that as we were out of sight of the
usual landing place for boats, that we might be
missed if we did not get into another situation.
Accordingly, next day, I set off with a guide
and some of the people to walk along shore, and
look out for another place where the tent could
be pitched near water and opposite Pelican
Point. At the distance of eight miles from the
kraal we observed through the mist which lay
on the sea, a white object on the water—we approached nearer—when old Choubib, seemingly

mad with joy, began to dance about, and to shout, " een schuit ! een schuit !" a boat ! a boat !

Presently an American whale boat under sail neared the beach with a shark in tow; the crew of which seeing, unexpectedly, such extraordinary figures as I and my men were with our beards, ostrich plumes, hunting frocks and arms, hesitated to land; but hearing us speak English they stepped ashore, and, in the usual cool New England way, they shook hands with us without saying a word, when one of them, a mate, said, " what gang do you belong to ?" Now, *gang* in Yankee phraseology does not mean, as with us, a gang of robbers, slaves, or convicts, but merely a company, so I, having travelled the States from the Chesapeake to old Kentuck, answered, without feeling annoyance, that we had come from the Cape of Good Hope so far, on an Expedition of Discovery; on which the mate, with a half incredulous look, answered, " oh, H—l !"

I asked what ship had anchored in the bay, and the mate said it was the Commodore Perry,

Hoborn master, from New England, "and there is my captain," pointing to a lusty man in his shirt sleeves, who had just landed from another boat.—The captain came up in a friendly way, and said he thought we were shipwrecked mariners, for he had never seen or heard of white men before in this section of Africa, said he should like to see our camp, " at the head of the navigation," and "guessed" that we should like to come on board, and eat a little ship biscuit and drink a glass of grog. I thanked him, and said that I was on the look out for a new "location" for my tent, but would be with him by-and-by. I then directed Taylor to accompany the captain in his boat on his way to the tent, and went with Robert, Elliot and the guide among the sand hills, to find another watering place.

We had been landing another cabaljao, and had left our shoes under charge of a boy. At first on the soft mud we felt no inconvenience, but when we got on the burning sands we longed for the receipt of the Fire King; however on we trudged, crossed the mouth of the Kuisip, in which there appeared to have been water only after floods in the river; toiled up and down the sand hills, and found in the different vallies between them six or eight holes, in which was brackish water; but we saw no place where the oxen could have found food, half so good as where they were, though the pasture there was coarse enough.

After labouring for four hours under a hot sun, sometimes half way up to the knee in sand, and with our feet scorched with the heat, stung with the quick grass, and bruised with the baked clay, we reached the tent, and found Captain Hoborn and his people there. We had nothing to offer the strangers but some 'naras fruit and brack water, which last the Americans could not swallow. I told them we had suffered so much from want of water of any kind, that sweet or brack, clear or muddy, was all the same to us, provided we got a belly full of it.

" That's d——d hard, I swear," cried the carpenter of the whaler.

" Can we get any green or fresh here?" (vegetables or fresh meat) was next asked.

" We have seen none yet," I said; " we are

ourselves much in want of provisions, and would be glad to trade with you for a little ship's beef and biscuit."

" What can you miss ?" was asked.

" Some rope, knives, sambuks or whips of rhinoceros hide, pipes, and zebra head skins for pouches."

" Well, come on board, and see what we have got, and speak to the Niggers here, will ye, for some *fresh* for us, and, we'll miss them a musket for two or three bullocks."

I went off with the Americans toward the boat " at the head of the navigation," and we found it high and dry, with the boat-keeper asleep in it. He was saluted with this strange abuse, " You've been taking a dodger, eh ! you damned h——l !" and we then put our shoulders to the gunwale of the boat, and shoved her over the mud into deep water again.

I was hospitably entertained on board the Commodore, and enjoyed especially the biscuit, potatoes, and penguin's eggs boiled hard, the yolk of which is capital eating. Capt. Hoborn said he thought of remaining four months at

Walvisch Bay, that he was now looking out for
Hunchback whales to come in every day to
breed, and that they had] already got some
fish lower down the coast.   They never heard of
any British whalers coming to Walvisch Bay,
but saw an English brig at Angra Piquena
lately, and said, that our people seemed to over-
look the fishing on the African coast almost
entirely, which is certainly true.*

I told the Americans several stories of the
chase in return for their good fare; and one of
the mates " calculated" that he would make his
fortune in a month if he had " that runner of
ours" (Henrick) in New York.

After a comfortable sleep in a berth, I shewed
the Americans where to obtain a large supply of
fire-wood at the mouth of the Kuisip—trees
brought down by the floods in the river; and
then I returned to the tent, where I found the
chief's wife waiting to see me.   She was an old
woman, lame of a leg, and attended by half-a-
dozen ancient ladies of honour, from whom she
was distinguished by wearing a handsome kaross.

* See Appendix.

of jackal's skins, and the handkerchief I had sent her for her head. I told her the object of my journey, that I was now looking out for a ship to get assistance, and in the meantime, I should be happy to barter handkerchiefs, beads, knives, &c. with her or her people for some cattle or sheep. She promised to do what she could for us; and after I had given her a few small presents, and above all, some tobacco, she went off in a good humour.

The Chief Kuisip coveting one of the ship's muskets, said he would give a couple of his riding oxen for it; and Henrick said he would "miss" an ox for five bottles of powder, (two is commonly given at Angra Piquena for an ox). I did not like parting with any of the cattle, not knowing but that we should be reduced to eat most of them yet, and abandon the baggage; but as Kuisip and Henrick had conducted themselves so well towards me, I did not throw any difficulties in the way of their bargains, and accordingly they and Choubib went on board.

Another whaler now appeared in the bay, the Pocahontas, Menter, from Portsmouth, United

States. This ship having been out longer than the Commodore, and having had no " green or fresh" for some time, was afflicted with scurvy, but which I saw cured in a simple and novel way. Capt. Menter got some potatoes from the Commodore, and bringing his patients on deck, he made them eat for three or four days a few raw potatoes, washed and sliced, and the effects of this treatment were astonishing—the men's gums, which before were white and sore with disease, resumed their natural colour, and the other symptoms of scurvy also left them.

The morning after the three headmen had gone on board, I was looking for fish on the beach, when I noticed Kuisip, Henrick, and Choubib returning to the tent, and every now and then looking, between the light, at a bottle they had got, and seemingly in high argument, I went to them, and found them a little " raised" with liquor, and in a great passion.

" Look, mynheer," said Choubib, " at the trick which has been played us by one of the mates. We got five bottles of what we thought was powder, but one of them we now find to have only a little fat in it."

I looked at it, and found that half a bottle of palm oil had been given to the Namaquas as a *bonne-bouche*, or to make their woolly hair grow, perhaps; but as they wanted powder and not pomatum, I took the oil from them, and promising to get the mate to rectify the mistake, I sent them off to the tent to keep them quiet, as they talked big of shooting, &c.

I respect the Americans as a nation for their stirring activity and steady perseverance to raise themselves in the world; but the respectable citizens of the Union must condemn the slim "tricks" which some of their people from particular sections are too apt "to play on travellers"—such as the one now attempted. I don't think Captain Hoborn knew any thing of it till I told him, when a bottle of powder was immediately supplied.

On the 29th of April, the first hunchback whale appeared in the bay, and an active pursuit took place immediately with half a dozen boats.

The American cedar boats, with the weight well forward, seemed to pull better than English boats. The whale was soon hemmed in, and we thought it was a prize, when, after rising and spouting for the last time, it disappeared with a bellow, dived under the boats, and carried out its great bulk to sea again.

On our first arrival at the bay the wind was often S.S.W., with thick fogs and small drizzling rain, so that the appearance of our encampment in the midst of a sombre plain, with some hills indistinctly seen about it, reminded me of a dreary scene in the arctic seas. Towards the end of April the wind chopped round to N.N.E. For three days we had a gale from the S.W., during which the thermometer was at 70° at noon, and we were now (half roasted as we had been formerly), quite benumbed with cold, and my Namaquas became impatient to leave the coast.

Through the kindness of Mr. Hayes, a fine young man, a mate of the Commodore Perry, I was twice landed, with three or four of my people, at Pelican Point, the best place to dig for clams. It was rather an odd employment to go down on one's knees as the tide was receding, and black shags and white gulls were screaming round one, and wingless penguins were shuffling along the beach of the dark main, and to dig with one's hands in the wet sand, and at half a foot under the surface, to find the desired shell fish. I have not much of "the kid glove or silver fork" in me; still this occupation rather spoilt my nails; but what will not one do for dear life—for food! We got bushels of clams at Pelican Point, and they ate very sweetly at the tent.

At last, after a good deal of negotiation, the Bay people, (who were now in considerable numbers, men, women, and children), brought some lean sheep and goats to barter. We exchanged beads and cutlery for them, and again made up a small flock. We also got for rope, fishing lines, &c., two or three bags of ship's biscuit, and Captain Menter, (a worthy kind hearted man), knowing our late sufferings, seeing our present state, and fearing that we might yet perish if we attempted to go further, offered to run over

to St. Helena with me and my seven Cape attendants for 70*l.* or the price of a whale; but I said
that I had not yet seen enough of the interior,
that I intended (since I could not go further to
the north from Walvisch Bay) to go as far east as
I could, and having now a small supply of food
for present support, I trusted ere long to find
game again. I thanked Captain Menter for his
offer of a passage in the Pocahontas, though I
never felt tempted to avail myself of it.

I now held a council with the headmen about
our further proceedings. Henrick Buys said he
would go with me to the world's end if I chose,
the determination of this fine fellow and prime
hunter was quite an " I pre, sequor" one, and I
highly appreciated his resolves and merits; as
to Kuisip he was also willing to assist me with
his own services and those of his people, but he
was under the guidance of the cunning old fox
Choubib, the interpreter, who seeing that the
man-of-war, from which he expected so much,
did not arrive, he did all he could to persuade me
to return by the shortest road to the Orange
river, pretending that he was quite alarmed

about my resolution to go to the eastward, that
we should now certainly perish either from hunger, thirst, or the wild Damaras, and that the
only chance he saw of saving the expedition, was
by returning to Ababies again, and recrossing
the Great Flat.

I answered that I would sooner die than again
see Calabash Kraal, where began our greatest
miseries; that if he did not choose to go further
with me, I should now give him the musket
and the other things which had been promised
him, and that he might depart in peace, but that
nothing could induce me to give up the attempt
to penetrate to the east. Seeing that he could
not carry into effect his own secret intention of
returning direct to his people, he at last said he
would also accompany me to the sources of the
Swakop and Kuisip.

Choubib was no favorite with my people: he
was a short tempered pragmatical old fellow, and
was also excessively greedy; he carried with
him sundry bags, in which he stowed away whatever odds and ends were lying about; small bits
of tobacco, straps, buckles, needles, buttons,

soap, &c., and it was very difficult to prevent a positive fight between him and my attendants, white and coloured, so captious and quarrelsome was he.

As we could have now done without him (having Henrick Buys, who could also interpret for me) I wished Choubib to leave for the sake of harmony; but he would not, and even tried to persuade me to remain longer at the sea for the man-of-war.   I had waited a fortnight and saw no signs of her, and we had had so much difficulty in getting a small supply of provisions, that I could not remain longer by the sea side, eating up every thing, and consequently was obliged to move.   Why Choubib was so glad when he first saw the whale boat, was because he thought that it belonged to the man-of-war I expected, and that if I sailed in her, to him would fall my bullocks, stores, &c. and that he would obtain besides, a handsome present from the vessel!

As we are about to leave Walvisch Bay, the question naturally arises, is it well adapted for the establishment of a religious mission, or of a factory for trade?   Besides Angra Piquena, it is the only bay on the south-west coast of Africa, of any size, until Saldanha Bay is reached. It is a very safe bay, the holding ground is good, nothing can hurt a vessel anchored behind Pelican Point, and there is plenty of (brackish) water, and of fire wood. It teems with fish and wild fowl, and must be a favourite resort for whales, or the American whalers, sometimes two or three together, would not remain here for four months as they do. The tribe which inhabits the shores of the bay is a large one, that is, some hundreds in number; for I saw many groups of their huts among the sand hills; and though a wild people, they might be conciliated with kindness. They have flocks and herds, though we saw few of them, and those only of the worst description; for they were doubtless afraid of tempting my Namaquas to make a foray amongst them on a future day. It might be worth while to ship cattle from Walvisch Bay to St. Helena. In the time of Napoleon they used to be sent from Benguela. Seven hundred ships put in annually to St. Helena, and cannot obtain there the supplies they want. One hundred and fifty or two hundred miles N.N.E. of the bay the country is full of fine cattle; and even the bay people can produce a good many from their sand hills, when they think there is no danger of shewing them. There is a possibility of much ivory being obtained at the bay; as further north the country is certainly full of elephants.

The climate of the bay is healthy and good. It is hot in the beginning of the year; but in May it was cool, and it would continue so till August. There is no stagnant water, and nothing to cause fevers about the bay. The great drawback to a settlement here would be the light and sandy nature of the soil. Yet it is astonishing what the pure sand of Africa produces with the addition of a few decayed leaves, and with moisture. The people said there was plenty of mist (or small rain) in the cool months, which would bring forward vegetables, though there is no stream which could be led out over the land. I sowed some melon and pumpkin seeds by a pool.

A Captain Morell, of the United States, said that, from what he saw of the people of Walvisch Bay, he was convinced that by placing himself under their care he might have gone right across Africa. The captain was quite wrong in his notions regarding the bay people. It is very unusual for them to go beyond the mouths of the Swakop and Kuisip. The chief has no influence beyond the shores of Walvisch Bay. No one can pass through the Damaras of the plains from the bay without a very powerful escort; and the only thing which might be done, (besides what we did), would be to induce the chief to show the way to the Red men living to the north, that is if they do not come down to these a coast in latitude 19 or 20, and could be more easily communicated with from the sea. This expedition would be well worth a trial. I had the greatest desire to undertake it; but, besides not being able to get any guides to go with me to shew me the waters, our cattle were now so feeble and knocked up, that it was doubtful if we could get many of them beyond the sand hills. Without guides then, and without some cattle in tolerable order, it was impossible for me to see a most interesting race; and how as red men they are nearer the line than the negro Damaras I cannot say, unless the Damaras came from the north east to the Swakop.

If missions were established farther in Great Namaqua land than the Warm Bath, it would be necessary to have a station at the bay, to assist and communicate with those in the interior. It would be too far to send to the Cape for supplies with waggons for stations about the sources of the Great Fish River, for instance; and therefore a bay station would be indispensible: and perhaps, with prudent management and caution, tempering zeal with knowledge, the fine race of the Damaras of the plains might be communicated with, and without danger, from the bay.

Our principal amusements at the bay were shooting wild fowl, (to keep the people from wearying), and eating 'naras and shell fish. Two or three times we hauled the seine, which, however, was rather short for sea-fishing, but we managed to catch mullet with it. I wished to go in a whale boat to the mouth of the Swakop, to ascertain the existence of elephants, which are said to be numerous about the river ; but some excuse was made for not lending the boat, and the bad plight of my bullocks prevented my going by land.   Once out in a whale boat, it blew very hard, and with a sail set, and steering merely by the trim, and without an oar out, we were on the eve of being upset, with a strong gust, among the hungry sharks, and were only saved by the mast going by the board.

The scenery about the mouth of the Swuakop was striking from the sea.   The sandhills, which extended from the Kuisip, were here succeeded by mountains apparently two or three thousand feet high, and called the Qua'nuas, or clay-bank-trap mountains, that is, those in which the foot is caught as in a trap: the sailors called them the Blue Mountains.   I named the highest of them Mount Colquhoun, after my valued friend and connexion, Gideon Colquhoun, Esq., late Resident at Bussorah.

CHAPTER V.

ON the 3rd of May, for the last time, I went
up a sand hill commanding a view of the long
line of coast, and the broad expanse of ocean;
but I saw no ship of war, or any vessel, save the
two whalers lying in the smooth water off Pelican
Point. Whilst about two or three miles from
me, and south of the bay, the waves beat with
hoarse and constant roar among breakers.

A considerable time was now spent in collecting the sheep we had purchased from the
bay people, and in bargaining with them for
some strings of copper beads, which they said
they had got from a man who lived on a hill
north of the Swakop. At last every thing was
packed, and in order for the march, which was
now under the guidance of two of. the bay
people, who promised to show us the waters in
the Kuisip, that we might not experience again
the same difficulties we had formerly felt in
coming down the river.

We drove the weary oxen, attenuated with the
salt grass and brackish water of the bay, up and
down the sand downs ; the packs came off some,
and others stuck fast; and we were obliged to carry

before us, on the oxen, some of the short-legged
goats of the bay, which could not keep up with
the flock. We dined off an old cow bought
from Choubib's brother, which was as tough as
sole leather; and in the evening we halted to
sleep under a sand hill. After five hours'
journey we were unable to go farther for the
heavy and wetting mists which enveloped us,
and which gave many of the people severe colds.

Next day we reached Red Bank. Here, with
reeds, sweet grass, and good water, the oxen
recovered a little, and the spirits of the party
were raised. An old man on a journey, and in
charge of eight women, here joined us, and said
that Quasip, the chief of the bay, was passing us
behind the sand hills, afraid to approach us. I
immediately sent off one of the guides to intercept him ; but he did not come to us till he had
been at the bay, and had questioned his people
there regarding us, and on the evening of the
6th he appeared with six followers.

Quasip was a cunning looking man of about
forty-five years of age. He sat by our fire
wrapped up in his kaross, peering warily round

'him, and with an old councillor at his ear. He had now come from the country of the Hill Damaras; and he reported that a bloody battle had just been fought between them and the Damaras of the plains, in which the latter had gained the advantage, and had massacred many of the women and children of the hill people. He also said that some distance up the Kuisip, we should fall in with plenty of rhinoceroses, and also obtain other game to support us. This was good news for us; for now I really thought that the Namaquas, in their desperate hunger, (whetted with scanty portions of the small sheep of the bay, and with the keen air of the sea), would have risen at night, and slain and eaten without leave asked or obtained; and that there might even be a chance of their eating one another before we got to the game. As for myself, I was tolerably safe from their devouring jaws, being in hard working condition, with little flesh on my bones to tempt them.

I told Quasip that he must spare us something to kill, and I made him a present of knives, tinder-boxes, &c., to encourage him to send us

something to stay our appetites. He ordered
one of his people to bring us two head of cattle;
and a fat young bull and a heifer appeared after
a few hours, which proved a most seasonable
supply for us.

I of course questioned Quasip regarding the
Nubees, or red people, to the north of the
bay, whom he had once visited; but I could get
no other information from him than this, that by
good luck he had passed by the Damaras of the
Swakop, had gone a month to the north of the
bay, and had there fallen in with the great red
nation, who were very friendly; spoke a different
dialect from the Namaquas, but that he understood them; and that they were distinguished by
allowing their woolly hair to grow long. I asked
him if he would go again to the Nubees. He
said that it was inconvenient for him to leave the
bay at present, having just now come off a
journey.

We got two Boschman guides from Quasip,
instead of the two bay-men. The new guides,
Oahap and Numeep, were to accompany us to
the Hill Damaras, and our minds were now set

at rest about the water even beyond the Kuisip. On the 8th we were again on our way up the river, and halted at Gnuhooas, or Black Hole, twenty-four miles from the Red Bank.

About two hours further on was the watering place of Gnutueip, or Black Nose. Here were the graves of the Damaras, who were pursued up the river and slain by the bay people, and here also we saw the last of the 'naras fruit.* At Hou'tous, or Sand Gate, we had a delightful " off pack" under shady trees, with plenty of good grass and water. Some of the oxen stuck in the mud here, and we saved the weakest of them with difficulty. A fresh lion spoor kept us on the alert, but the Namaquas would not consent to go after him, because he had spared us. The people here roasted, pounded, and ate all the pieces of ox hide we had left: having suffered thirst in coming down the river, we now endured hunger in going up.

* Some plants of 'Naras are now growing in England (March 1838) from seeds which I brought home; they are a foot high and beginning to branch, having two thorns at each articulation, and a stipule scarcely to be called a leaf between them, on the axis of which is the bud, but no leaves.

It is a custom with the Namaquas when in the field, and ignorant of the water places, to look about for Boschman spoor, to catch one of the "children of the desert," and to make him show where water is to be found.   On one occasion two or three Namaquas were returning from the coast, and unacquainted with the pools, they were dying of thirst, when they fortunately fell in with a Boschman.   "Where is the water?" they inquired.

"I don't know of any water here about," said the Boschman.

"What! are we to die here? Come! take this stick and dig here in the ground for water."

The Boschman, in fear of his life, did as he was bid, and for a short time he turned up the dry surface of the plain, but soon tiring of work, he stopped and said, "Oh! I just remember there is water over the hill."

To avoid a considerable southerly bend in the Kuisip, we now left the river, and stretched across a hard gritty plain. Seven miles from Hou'tous, I found two cameleons, crouching on the ground beside some stones, and not within miles of either bushes or water; their haunt seemed to be the open plain, unlike common cameleons, which are always found on or near bushes. The occiput of these new cameleons is triangular pyramidic, with the keels slightly denticulated; on the back is a series of distant, rather triangular, and blunt tubercles. The scales of the body are small, and nearly uniform, while those of the head are flattish. The tail is cylindrical. When I approached these cameleons, named *tuberculiferæ* by Mr. Gray, of the British Museum, the little things opened their mouths at me, and hissed like angry snakes, whilst a bag under their mouth swelled out to a great size, which, with their dark blotched bodies, gave them a hideous appearance. They run fast, and are accounted to be poisonous by the natives.*

At the distance of thirty miles from Hou'tous, we passed a grotesque collection of rocks, rising with dark and vertical stratification and serrated edges from the broad plain. The rocks were called by our guides Einhiras, or the Hill of the

* See Vignette.

Laughing Hyena. " Here," said the Boschmans, " is found a most extraordinary snake:
eight feet long, mottled back, with overhanging
brows like a man, and fiery eyes; whilst the sex
is as plainly distinguished as in beasts. It lies
commonly stretched out under the rocks, and we
are much afraid of it."

I wished much to have halted a day at Einhiras, to obtain a specimen of this strange reptile,
but there was not a drop of water near it, so we
were obliged to hurry on in the night, and off
packed among some bushes, after a thirty-five
miles march. On the 12th, we passed on our
left Tarahap, or Quiver Mountain, and Hokap,
or Spotted Mountain, and saw before us the
range called Tumas, or the Mountains of the
Wilderness. Their height, like that of Tarahap
and Hokap, seemed to be about 1800 feet.

The morning scene recalled to mind this
quotation,

What dreadful pleasure ! there to stand sublime
Like shipwrecked mariner on desert coast!
And view the enormous waste, of vapour tost;
In billows lengthening to the horizon round,
Now scoop'd in gulfs, with mountains now emboss'd.

After twenty-eight miles, we came to the Humaris, or Rolling river, into the deep rocky bed of which we descended with difficulty, and found ourselves at the bottom surrounded by precipices of two or three hundred feet high. The guides searched about for water holes, and, at last, far under a rock, a cupful of the precious element was found for forty thirsty people and seventy parched oxen. A party stripped to the skin, and as diggers relieved one another to clear out the hole under the rock, and the people all drank sufficient, but the oxen only obtained two or three fillings each of a large metal dish we had for meat;—the poor things were much distressed. We slept close under a huge mass of granite, and expected a visit from a lion, whose fresh footmark was beside us—but he did not trouble us.

Next day, we passed along the bed of the Humaris, enclosed all the way with lofty cliffs, between which the heat was so great that I thought I should have got a stroke of the sun. We walked on foot, and rode the oxen occasionally. Saw numerous traces of rhinoceroses, but the animals themselves trotted away out of sight, up the rocky glens which led into the river. A pursuit took place after a troop of zebras on the top of the bank, and one of them was shot.

Our course was E.S.E. We passed two pools of water, and had to slip down two or three dangerous steps of rocks which ran across the river's bed. The oxen, loaded and unloaded, had to bring their hind legs under them, and to slide down the smooth inclined plains, and we got along without any serious accident. At last we saw, in an opening between the rocks, the very refreshing sight of green trees, grass, and water; this was in the bed of our old friend the Kuisip. We gladly off-packed, after seventeen miles, under a mighty camel thorn, opposite the junction of the two rivers, and where we saw a lion had been in close pursuit after some baboons, who had escaped before him among Palma Christi, or castor oil plants.

Our herd of oxen had not seen more grass than would have satisfied one of them, since we had left the Kuisip at Hou'tous, three days before, and eighty miles distant; they now rioted in plenty; and we bipeds also felt very comfortable and happy with our respite from fatigue and suffering. One of the men played "the Gorah's humming reed," which gives out wild tones like those of the Æolian harp, or of a distant horn. The instrument consists of a bow, part of the length of the string of which is a slip of ostrich quill; and this, being applied to the lips, gives out the melancholy sounds which so delight the tenants of the desert. Besides the meat of the zebra, I ordered our fattest heifer to be killed, and the people laid down to sleep, satisfied with abundance once more. We were indeed as well off as poor people could be in the wilderness, and I felt free from anxiety about water and game, for some distance at least, seeing that we had two capital Boschman guides with us.

After allowing Oahap, the elder of the two, to occupy himself for some time about the stomach of the heifer, I called to him to come and speak with me by the fire. He rose reluctantly, with a handful of bowels, and came and sat opposite me in no pleasant humour; but a pipefull of tobacco soon smoothed his brow.

"Have you," I said, " always lived about the Kuisip?"

"Yes."

" And your father before you ? "

" Yes."

" What was he ? "

" He was a great chief: the people under him stood like the reeds over the river, they were so many."

" What became of them ? "

" Some were destroyed by wild beasts—as the lion, the elephant, and the rhinoceros; others were killed by the Damaras; and the rest died from hunger and old age. Of my people there are but few left now."

" Think now, and tell me, " I continued, " what is the most wonderful thing you ever saw in your life ? "

Oahap was sorely puzzled at this question. His range of observation had not extended far. He had apparently lived about sixty years in the world; he was tall and stout, still vigorous and active; and his " beat " had perhaps never extended beyond the branches of the River of Roots.

" The strangest thing I ever saw," said he, after a long inspiration, and swallowing the

smoke, " was this.   One day, two of us found
the fresh marks of a couple of rhinoceroses in
a path.   We made a little stone kraal by the
side of the path, where my companion lay in
wait with two assegaes in his hand, and I went
off to look for the rhinoceroses, and to disturb
them.   I found them asleep some distance off,
under the trees: one of them was an old cow,
and the other a large calf-rhinoceros. I threw a
stone, they stood up; I threw another, they
looked round; and seeing me, the old one rushed
at me in a great rage.   I ran off to a tree; and
had just got my feet off the ground in climbing
it, when the rhinoceros drove her horns between
my legs into the trunk of the tree; but I was not
hurt.   She then went off with her calf; they
passed the kraal; and my companion, standing
up, threw an assegae at the old one: she went a
little way and fell.   He stood up again, and
threw the other assegae, when the calf also fell
dead.   I came on after the rhinoceroses, and
seeing them both lying near the kraal, I jumped
on the back of the big one and rode it for joy,
and I cried out to my friend, ' Now I see you

are your father's son this day!' This, then, is
the most wonderful thing I ever saw."

Here we killed the first pheasants we had
seen: they are grey and brown speckled, and
were excellent eating. This variety is new, is
afterwards described, and we found them running
in considerable flocks under the bushes and
grass of the river's bank.

The Kuisip was here enclosed with hills; but
the crags were not so impending, nor was the
appearance of the river so forbidding as when we
had first seen it at the Devil's Den. On the
contrary, it was, at the junction of the Humaris,
picturesque and agreeable.

On the 15th we packed up, and ascending, by
a zebra path, the hills on the south side of the
river, we travelled in an easterly direction, and
up and down ravines, over much broken ground,
and round immense masses of calcareous soil,
which had slid down into the dry water courses;
and, after four hours' hard work, we packed off in
a hollow at Keree Kama, or jackal's water.

We sat at our evening meal in the bed of the
small river; and a strange discussion took place

between old Choubib, Henrick Buys, &c., about the difference of colour in the human race—the difference of language—the creation of man, &c. Of this last, some of the Namaquas entertain this notion—that the Deity having created white men, the devil became envious, seeing what a wonderful and handsome creature had been formed; and he also tried his hand at making a man; but he could not make him otherwise than black like himself, so in a rage he struck his man a blow on the face, which flattened his nose: and hence the negro colour and feature !

Whilst these knotty points were under discussion, an alarm arose that a rhinoceros was bearing down upon us, and so it was. The monster was steering down the confined bed of the Keree Kama, for the small water place beside which we were sitting in the sand. Every one was on his feet in a moment; the meat dish was upset, and the precious " tea water" spilt in the sand, in the hurry to scramble up the rocks. But before we could get a shot at the rhinoceros, he " turned tail" and disappeared.

We were now in the country of the Damaras

of the Hills ; and we saw over the broken ground we traversed on the 16th, single mountains rising in the most grotesque forms, serrated, and peaked, to the skies. We also crossed the fresh traces of several lions, and then reached a great plain surrounded with mountains. On the plain we saw in every direction zebras grazing in herds of six or eight. I had never seen before such a number of these beautiful animals together; we seemed to have got to their head quarters here; and we were not long in extending our files, and securing some zebra flesh for our supper.

At the extremity of the plain, on a rising ground under a hill, I saw the first Damara village, but no inhabitants. The huts were of a conical form, and were composed of stakes meeting at top, and covered with grass. Round the bottom outside were placed stones to keep the grass from being blown away. To some of the huts there was a sort of porch to exclude the wind. Each hut was about ten feet high; and the whole eighteen were arranged at some distance from each other, in a circle. In the middle was the dancing place; but there were no kraals for sheep or cattle.

The village was a mile distant from the water (a pool of the small river Numsep, or man's-kaross-lay-aside,) that the wild beasts might not be disturbed in their passage to the water by the vicinity of men. The hill above the village is a place of retreat; and it is the custom of the Damaras to sound an alarm, upon the sight of strangers, with a cow's or deer's horn, and to run up the hill to defend themselves, if necessary, with arrows and stones.

About the Numsep the number of rhinoceroses was very great; old and fresh traces were seen every where; and after a twenty miles' march, and packing off at the pool called Onakusis (or woman's-petticoat-water), with the mountains called Oosip (joining,) and 'Nabagno (or rhinoceros horn) to the north of us, the chase commenced, and continued for two days. Several rhinoceroses were wounded but not secured; but two zebras and a gemsbok were added to our

larder: whilst flocks of blue pigeons, night partridges, and even parrots frequenting the pool,
gave us fowl as well as flesh to cook.

On the 18th we travelled E.S.E., and crossing
a ridge, got into the bed of the Kuisip, or dung
river, which was well lined with bushes, and full
of rhinoceroses.

During the journey I had often endeavoured
to find out traces of religion among the Boschmans and others; but I had hitherto been very
unsuccessful.   I have before alluded to the superstition of Heije Eibib, among the Namaquas;
but among the Boschmans I had discovered
nothing to indicate the faintest trace of religion,
but now I did in a singular way.

We proceeded up the Kuisip, and among the
grass we had excellent sport with numerous flocks
of guinea fowl, which we had not seen since we
left Habunap; and after thirteen miles, we
packed off at two deep holes in a rock, full of
excellent water, at a place called Abashouap, or
" red man's child," when 'Numeep, the Boschman guide, came to me labouring under an attack
of dysentery, and said that he was about to die !

G 2

I asked him what had occasioned the disease; and he said it was from having dug for water at the place called Kuisip, in the bed of the Kuisip River, near our last watering place, *without having first made an offering*, and that therefore he was sure to die unless I could help him. I gave him immediately two table spoonsful of olive oil on water, which relieved him—(I here beg to advise that no traveller be without this invaluable remedy for bowel complaints)—and I then asked him what he meant by saying that he had made no offering at Kuisip.

"Before any Boschman," said 'Numeep, "digs for water at Kuisip, he must lay down a piece of flesh, seeds of the 'naras fruit, or an arrow, or any thing else he may have about him and can spare, as an offering to Toosip, the old man of the water."

Now on this occasion 'Numeep had left nothing at the water, and was therefore afflicted for his neglect.

I asked 'Numeep if he had ever seen Toosip.

"No; I have never seen him, nor has any body else that I know of; but we believe that he

is a great red man with white hair, and who can
do us good and harm. He has neither bow nor
assegae, nor has he a wife."

"Do you say any thing to him when you put
down your offering at the water-place?"

"We say, 'Oh! great father! son of a Boschman—give me food; give me the flesh of the
rhinoceros, of the gemsbok, of the wild horse, or
what I require to have.' But I was in such a
hurry to drink this morning, that I scratched
away the sand above the water, and took no
notice of Toosip; and he was so angry, that if
you had not helped me I must have died."

Having indulged too freely in zebra flesh at
the last water, was doubtless the cause of
'Numeep's illness; but fear may have made him
worse. I was very glad he had been ill; for
owing to this, I found out a trace of worship
among a very wild people.

We now saw miles of hedges, about three feet
high, laid to direct the wild animals to pit-falls
placed here and there for them; the pit-falls for
the rhinoceros were four feet deep and four broad,
with branches and leaves over them, and were

consequently not large enough to take in his whole bulk, but were only sufficient for his fore legs, which the people said was the best way of securing him, as his legs once in, they have no purchase with which to raise his body. There were also other means for securing the smaller game. Thus a cord formed of the inner bark of a tree was tied to a young sapling, a loop was made in the cord, and the sapling was bent down and fixed slightly to two cross sticks; the loop was opened and arranged on the ground above a hollow place and under a few blades of grass to conceal it, so that a deer, or even an ostrich, on passing through the opening where the noose was placed, and putting a foot through it, was immediately twitched into the air by one leg, and thus became the prey of the Damaras.

I ascended an eminence above Abashouap, and was much struck with the grandeur and beauty of this part of Damara Land. Looking towards the east, and at the distance of eight or ten miles, rose the huge mass of the 'Tans mountain, with its square top and furrowed sides; lesser heights were beside him, whilst the whole country was a series of ridges and valleys, on which were scattered dwarf trees and bushes, whilst fine grass waved gently in the breeze in every direction.

Huts, three and four together, of the same construction as I had lately seen, were observed in many parts of the varied and extensive landscape; but I did not see a human being. The guides said that last year there had been a drought and famine in the land, many of the Damaras had died of hunger, and the others had moved off for a time to the eastward, where more rain had fallen.

Two rhinoceroses, an old dam and her weaned calf, were observed lying asleep under short and stout trees in a valley near Abashouap. They were cautiously " becrept," and the old one was shot. All night a party remained by it to cut and " vlek " the meat, for carrying off a quantity of it; and the young rhinoceros alarmed them by coming close to them in the night to look for its mother.

We had now in our pot at one time the flesh of the rhinoceros, zebra, gemsbok, and hare, also

guinea fowl and pigeon; but we had no biscuit
or vegetables to render this variety palateable.
One thing I now remarked, that after partaking
of rhinoceros soup I was much stronger in walking and running than at other times; but the
flesh of the rhinoceros is coarse and rank, and
only does for a "bush appetite."

· On the 19th we travelled S.E., and a large
black snake, ten feet long, was seen steering
towards some rocks with a hare in its mouth.
The guides were disturbed at seeing this snake.
"That is the komakasip," (or what-cannot-bear-
the-sight-of-cattle), said they. "It is the most
dangerous of the snakes in this land. A man
runs but a short distance after he is bitten by the
komakasip. Some time ago a Boschman discovered a honey nest not far from his hut, and
he was creeping into the hole to rob the bees,
when a komakasip bit him in the face. He ran
home as fast as he could; but he fell dead
before his own door."

We shot now some new birds about the size
of a thrush, with blueish backs and yellow
breasts: they are afterwards described. We

also got a small variety of magpie, black with white wings, and finches with red heads and speckled breasts. After nine miles of hill and dale, we packed off in a deep dell, at a place called Unus, or narrow river, with 'Tans towering four thousand feet above us on our left.

The cattle had fortunately become stronger within the last few days, with the sweet grass and the good water, and we were thus able to climb the steep and pathless offsets from 'Tans, (which are called Kumap, or the mountains of supplication), though the fatigue of doing so was very considerable. After five miles of a winding course on the 20th, we breakfasted at Eisees (or beautiful) Fountain, and then surmounting a very steep ascent, we zig-zagged down on the other side, and saw deep water courses still further below us, and a solitary eagle floating above us.

The perspiration poured in streams down the people, and the cattle were white with foam, when we got to the bottom of a valley, and commenced another ascent, steeper and more rugged than the last; but at length, after desperate efforts, we got all the cattle on a table land, when the cool air refreshed us, and the magnificent prospect of many miles of mountain scenery towards Ababies, and the desert where we had nearly perished.

There was one consolation we had under all our fatigues, that no white man had ever before traversed the scenes amongst which we now toiled.

After an easy descent, and at the distance of thirteen miles from Eisees, we came to Chuntop, or sand path, a beautiful place under rocks, with high trees, and grass up to one's waist. The poor cattle roamed about among this, and were confounded with abundance.

> " As in the storm that pours destruction round
>   Is here and there a ship in safety found ;
>   So, in the storms of life, some days appear
>   More blest and bright for the preceding fear ;
>   These times of pleasure that in life arise
>   Like spots in deserts, that delight, surprise,
>   And to our wearied senses give the more,
>   For all the waste behind us and before."

DAMARA WARRIOR.

Published by Henry Colburn Great Marlborough Street, 1838.

CHAPTER VI.

I now learned from my guides that we were in the immediate vicinity of a Heis, or Damara village, which had not been deserted, as the others had been, which we had already past. I had not allowed any firing for four and twenty hours, as I was afraid of alarming the Damaras, and I now sent off Oahap alone, to prepare the Damaras for our seeing them, and to assure them that they should not be harmed in any way.

On the evening of the 18th of May, Oahap came down the glen, at the bottom of which we lay, with the head man of the Heis, and three others, at which I was much pleased.

These Hill Damaras were about five feet seven inches in height, and in colour and feature had all the characteristics of the negro, even to the projecting shin bone. They came with long staves in their hands, and without arms, in token of friendship and confidence, though perhaps their weapons were not far off. Their hair was peculiar; that is, it was cut off quite round the head, and an inch above the ear, leaving only the hair on the top of the head—in the manner of the Roundheads of the Cromwellian period. They wore short karosses of deer skin, and softened flaps of skin before and behind, to cover their nakedness; and in the hind flap, which was longer than the fore one, there was a pocket for holding roots, &c. They wore soles or sandals.

The head man was about forty-five years of age, and was a pleasant and communicative person. He said he would make one of his men guide us to the next village, Oahap and 'Numeep having fulfilled their bargain in bringing us among the Hill Damaras.

I asked the head man how he lived at this season, and he answered, " Badly enough. We are now eating mice, lizards,* roots, and sometimes leaves."

I inquired if he had always lived where he now did, and he said, " We have always lived among these hills; and we never knew of any other land."

I asked if he had any thing to do with the Damaras of the Plains. " No, nothing," said the head man, " they are our enemies; they are black like ourselves; but they speak a different language; we speak the language of the Namaquas; and the Damaras of the Plains, or Kamaka Damap (Cattle Damaras) speak a language of their own."

* The body of which has brown and yellow cross bars on it, and the tail is deeply serrated at the sides.

I told him that he and his people must not be frightened at white men, and that I intended visiting his village next morning. He answered, " Though we never have seen white men before, yet we always expected to do so. We heard always that they would one day come into the land, and we now see these strange men. I shall tell my people not to run away to-morrow."

Still adhering to the principle of not pitching the tent, when we were on the move, for the reasons before stated, we were rather inconvenienced by rain during the night at Chuntop; but as it was the first time we had heard rain drops for two months, the sound was not disagreeable, and now being amongst the fantastic rocks, grassy hills, and spreading trees of Damara land, we thought we should oftener experience refreshing showers than we had hitherto done.

We packed at sunrise, and the cattle going by a circuitous but easier path, north and east, I walked up the glen with a few men and riding oxen to the Damara village. We found it in a small mountain valley, surrounded with granitic rocks, amongst which were trees and shrubs, and with a citadel hill close at hand to retire to, on occasions of alarm. The twelve conical huts were arranged in a circle, and we now saw Damara women as well as men.

The women had their hair cut in the same way as the men, and many of them had lost two joints of one of their little fingers, which they said they had got cut off when they themselves had been sick, or their children had been ill. Cowrie shells hung from their heads, and half way down their faces. They wore short karosses on the shoulders, and over the fore flap or apron there were hanging short thongs, on which were strung pieces of reed, bones of hares, beads, blue and white stones, &c. The hind flap, like the men's, was provided with a pocket, for what the Dutch call "veld kost," country food, as bulbs, the fruit of the mysembryanthemum, &c.

By the doors of the huts lay bows and arrows, like those of the Namaquas; and in the grassy covering of the huts was stuck the usual throwing assegae. Clay cooking pots of a conical shape were in every hut.

The Hill Damaras are a numerous nation,

extending from the heights south of the Swakop to the Little Koanquip river, and they live in small communities under head men, in the manner we now saw them doing, without one supreme or paramount chief of the nation. They are commonly called Koup Damap, or Dung Damaras, by way of reproach by the Namaquas, whilst the Namaquas themselves bear a similar contemptuous epithet, among their constant foes, the Damaras of the Plains. I think 'Humi or Hill Damaras is the best term for the people with whom we had now to deal.

" We call them Koup Damap," said a Namaqua, " because they keep nothing to kill, and not even dogs to catch the fauns of the springbok, as the Boschmans do."

As the Hill Damaras have no cattle to transport mat huts from one place to another (in the manner of the Namaquas), their huts are permanent, and last for a long time; and sometimes they are covered with bark instead of grass.

The Hill Damaras cultivate no grain, only sometimes raise a little tobacco.

Few people are more simple in their habits

than the Hill Damaras, and among them there are hardly any ceremonies on those occasions when most other nations show marked peculiarities. Thus, when a man wishes to marry a girl, he goes to the father, with a present of bulbs and striped mice, to feast the old gentleman; and if he is accepted as a son-in-law, he adds to *the onions and mice,* an assegae or two, bows and arrows, a couple of karosses of springbok or rabbit skins, &c. and some of which he gets back again. They then dance a little (they make no honey beer at a marriage), and the bridegroom carries off his wife to his own hut.

Among primitive folks, like the Damaras, none live in single *blessedness.*

" Happy and free are the married man's reveries,
  Cheerily merrily passes his life,
He knows not the bachelor's revelries, develries,
  Caress'd by and bless'd by his children and wife,
Thus is each day one lovely holiday;
  Not so the bachelor lonely, deprest,
No gentle one near him, no home to endear him,
  No sorrow to cheer him, no friend if no guest."

The dance of the 'Humi Damap is somewhat similar to that which I had seen among the

Boschmans.   The women stand in a row, clapping their hands and singing, " Hey, he heyho ! hey he hey ! ho hoo !" whilst the men, with their sandals in their hands and with springbok's horns bound on their foreheads (which give them a Satanic appearance) stamp and dance round slowly before the women, and grunt in chorus.

The Damaras play on the gorah, which is their only musical instrument.   The Hill Damaras do not practise circumcision, as is the custom among the Kamaka Damap, and Caffres. In this respect the Humi Damap are like the Namaquas.

If a woman happens to curse or abuse her husband, they cannot sleep together any more; and the woman must then " eat from her own hand," or support herself—but they seldom curse.   If a woman goes into the field to search for bulbs, she never tastes them till her husband has first eaten of them.   In cases of adultery the adulterer is killed, and the woman is severely flogged by her husband.   The Hill Damaras take unto themselves as many wives as they can maintain.

A young Damara doctor showed me the way he cured his patients, and it was laughable enough. He provided himself with a clean wooden milk vessel, or bambus, and applying it, covered with a piece of skin, to the breast of a man who was lying on his side and groaning as if sick, he (the doctor) then left him, and sitting down opposite a stone, he began to strike it with the stick of his fox-tail *handkerchief*, and to sing at the same time, "To, to, to, tehei; to, to, to, tehei." After which he got up and danced round, and looked as if for something on the ground, at last he stopped suddenly, and appeared to find what he sought, and calling out "het, het," sharply, he goes to the bambus, and taking it from the patient's chest, on which he blows, he pretends to find some blood, or grease, or a bone in the bambus, which had been introduced by sleight of hand. The bambus is then carefully covered over, the doctor runs off with it a little way, and buries what he pretends to have conjured from the patient, in the sand, and then stamps over it, and the sick man is now supposed to be cured!

On the death of a person, a pit is dug, and the

corpse is placed in it in an upright position, and stones, bushes, and earth are placed about and over it, to prevent the wild dogs, wolves, or crows, eating the body.

Notwithstanding that some people maintain that there is no nation on earth without religion in some form, however faintly it may be traced in their minds, yet, after much and diligent inquiry, I could not discover the slightest feeling of devotion towards a higher and an invisible power among the Hill Damaras; neither had they any fear of an evil influence.

They believe in nothing but what they see. "Who gives you your food?" I asked.

"We get our living from the air—from the seasons," answered an old Damara.

"Why dont you keep sheep or goats, that you might live better than you do?"

"We have been afraid of losing them; we wished to keep them, but we thought the Boschmans would rob us of them. Now we think ourselves strong enough to defend ourselves and our property against the Boschmans, and we must try and get flocks."

" When you die, what becomes of you?"

" When we die we are buried, and are then no better than the beasts."

" Are you afraid to die?"

" Yes, very much; and we are afraid when we see people ill, because we think it may be our turn next—we try not to think of dying."

" Who do you think made the sun and moon, and all you see about you in the world?"

" We dont know; we are a stupid people; we never think of this. What is the use of thinking of it? no one ever told us any thing about these things, and how could we know any thing about them; all we want to know is, where to get a large animal to kill and eat."

" Do you, on any occasion, go to any particular place and make an offering there? For instance, do you go to a heap of stones and throw a stone on the heap; or put a branch on it; or leave a bit of skin on a bush anywhere?"

" No, we never do these things; we are a stupid people; we dont know or do any thing but look for food, and dance when we have got plenty."

I have given this conversation held with a Damara apparently as intelligent as the generality of the natives; and I think from it there is evidence sufficient to prove, that beyond their daily wants, the Damaras have no thought of any thing else; and " that," as Choubib, the interpreter, said, " they believe in neither God nor devil."

The mind of this ignorant people is like a " tabula rasa," ready to receive any impression, good or bad. There are apparently no superstitious notions among them to overthrow, no idol worship, no bowing down to stocks or stones. That they may at no distant day bow down to the true God, and that their minds may be instructed, and their spiritual and temporal condition improved, ought to be the earnest prayer and the endeavour of every lover of his species, who has the means to assist them.

I left the village, and with an old Damara guide, who was as fleet as a hare, we passed rapidly through grassy vallies, to intercept the pack-oxen, and the people with them. We were

now to the eastward of 'Tans, and we had striking
views of this noble mountain.

" Is there much game in this field ?" I asked
the old Damara.

" None," said he, when immediately after, to
give the lie to this assertion, a large troop of
white legged zebras, with sleek coats shining in
the sun, galloped across the plain, pursued by
some of our hunters, and we saw besides many
traces of other wild animals here.

The " trek," or pack-oxen, now appeared, and
from one of the drivers we heard of a loss which
I had just sustained. The driver was entrusted
with one of the boarding pikes, and seeing a
pack loose, he gave his pike to a Damara at his
elbow, to hold. After the pack was arranged,
the driver looked about for the pike, and he saw
the Damara making the best of his way with it
up a hill above the party, and deaf to cries
and threats, he disappeared over the top, thus
affording us an example of Damara roguery.
The quantity of iron about a boarding pike was
a temptation for a Damara which he could not
possibly resist.

We were now on a table land stretching from 'Tans eastward, one of the great steppes of South Africa, and the thermometer was 65°, in the middle of the day. On our right was one of the principal sources of the 'Oup or Fish river, round the head of which we were now journeying. We passed along with ease and comfort over a level surface, and arrived at the Taop or Cragless river, a branch of the Kuisip; here we found food and water for the oxen.

A number of Boschmans, tall and stout, (as those to the north of the mouth of the Great River usually are,) now visited us, and said their heis was a short distance off. I went a mile and a half and found twenty huts of stakes and bushes in a hollow. On some of the huts lay skins of antelopes of various kinds, which had not yet been prepared for clothing. I saw also some strange horns like those of a small cow, a mane, tail, and hoofs of an animal I had not yet seen; but I found out afterwards that these were parts of a Kaop, (master) buck or brindled gnu.

This is a very remarkable variety of antelope. It is about the size of a galloway; of a brown colour, with dark streaks over the body. The withers rise much higher than those of any horse, but the shape of the neck and body is somewhat equine. It has a long black mane above and below the neck, and a switch tail of the same hue. The horns are like those of the buffalo, but much smaller, and lie across the top of the forehead, then curve outward and upwards.

The Kaop is not found in this district in herds; they are oftenest found singly, or at most two or three together. It is a bold and resolute animal, and it is very dangerous when wounded, hence its name of " Master."

I was anxious to know how the Boschmans managed to kill the Kaop; and remarking two light frames covered with ostrich feathers, grey and black, on a tree, I asked them, through my chasseur Henrick, what they were. The Boschmans said, " with these we disguise ourselves as ostriches, and thus get near the Kaop, to shoot it with our arrows."

A present of tobacco induced a Boschman to disguise himself. He placed one of the feather frames on his shoulders and secured it about his neck; then taking from a bush the head and neck of an ostrich, through which a stick was thrust, he went out a little way from the huts with a bow and arrow in his left hand, and pretending to approach a Kaop, he pecked at the tops of the bushes in the manner of an ostrich, and occasionally rubbed the head against the false body, as the ostrich ever and anon does to get rid of flies.   At a little distance, and sideways, the general appearance of the Boschman, was like that of "the giant bird," though a front view betrayed the whole of the human body.   Approaching sufficiently near to the Kaop, which of course has nothing to dread from its feathered companion of the plains, the Boschman slips the ostrich head between his neck and the frame, and cautiously taking aim, discharges his arrow at the deceived Kaop.*

I was much amused with this manner of approaching game, and I contrasted it with the laughable way the Persians have of getting near partridges and quails, with a second pair of

* See end of chapter.

trousers rising from the shoulders, so as to make a strange figure with four legs (a pair of which are carried in the air), and thus confounding the birds with its novelty. Here the game were deluded by the approach of an object which was familiar to them.

But the Boschmans sometimes suffer when thus disguised. One approached an ostrich with a feather frame, and wounded it, when the bird ran at the disguised Boschman, and with its terrible toe-nail ripped him open from the breast downwards, and killed him on the spot.

The Boschmans of the Taop intimated that they wanted to come and dance at our " off-pack place" in the evening. Accordingly, a few of the men, and two or three dozen of the women came, and after I had lain down, they, with the assistance of our Namaquas, made a large fire, and the usual " Oh! ei oh! ei oh!" song, and clapping of hands commenced, which kept me awake for some time. In the morning, I found that the Bosch-people here had been as little scrupulous as those at Ababies, and all for the sake of a little of the zebra flesh we had.

At sunrise it was bitter cold, the thermometer indicating 40°, with a cutting easterly wind.

On the 22d we continued our journey, on one of the most disagreeable mornings for cold I almost ever experienced, though I had spent a winter in Russia. But there, fur-covered, one sets a temperature of 40° below freezing, at defiance; now we were not so well prepared for great cold, particularly those of the party whose raiment was scanty, or torn with the bushes in hunting.

Burns thus sings of certain comforts which we had no chance of obtaining for some time—

> " Fortune! if thou'll but gie me still
>   *Hale brecks*, a scone, and whiskey gill
> And routh o' rhyme to rave at will,
>   Take a' the rest,
> And deal 't about as thy blind skill
>   Directs thee best.

A few days before, I dreaded a stroke of the sun from the intense heat in the bed of the Humaris, now the blood was frozen in the veins. These extremes are sufficiently trying for the health, and have a tendency, in conjunction with hard work, to make a traveller older in appearance, than he is in years; but that is a trifle, if his constitution is still sound, and he has not been idly employed.

We passed over a most beautiful grassy plain, with scattered bushes and sand heaps, and on it we saw two or three rhinoceroses at a distance, but our fingers were so benumbed that we could not have pulled a trigger. The Karoo Koran, or small red bustard, flew up here and there to tempt us, but the cold took sporting out of us, till towards midday, when we saw gazing at us among some dwarf trees, a brindled gnu. This immediately fired us. The gnu shook his black mane and pawed the ground impatiently; we ran and crept towards him, but it was all in vain. He switched his tail at us, and went off at a hand gallop, presenting the appearance of a horned horse. Two or three others were seen, but we were equally unsuccessful with them.

We halted, after twelve miles, to cook and eat, beside a pool, and saw on our right the group called Kobip, or the Bone Hills. Continuing our journey on foot, for the cold, after sixteen more miles, we off-packed in the dry bed of the Chunchuap (or Hare Hole) River. Next morning, after an uncomfortable night from cold, we breakfasted at Chama, or Soft River, four miles. After which we saw a huge white or cream-coloured rhinoceros, on a hill, which moved about impatiently as the hunters ran up towards it. It seemed a mountain of flesh, and was, apparently, upwards of seven feet high. It went off with a ball in its neck.

The proportions of the head of the white rhinoceros are different from those of the black. The mouth is square; and the foremost horn is always, I believe, much longer in the white than the hind one. The fore-horn of the white specimen we had just seen, seemed to be between three and four feet long; and Henrick the hunter said, he had seen them up to one's shoulder. The white rhinoceros eats grass, and is a timid animal compared with the savage black species, which commonly charges, whether wounded or not; whereas the white variety tries to effect its escape.

On our left were the very picturesque mountains about two thousand feet high, called Aantup and Uep, or the Bird Stone and White Mountains; and twenty-six miles brought us with Boschmen guides, to our surprise, to a large Heis or village of Namaquas, called Naraes (or fallover,) on the Oanop, or Tell-tale River.

The Namaquas of Naraes were part of the people of the powerful chief Aramap, who lay still further in advance, and who had lately driven the Damaras of the plains from the beautiful and abundant country we now saw, beyond the Swakop. The Cattle-Damaras had, of late years, encroached greatly on the old Namaquas of the Upper Fish River, and were driving them before them down the river, when the conquered, being unprovided with guns, called on Aramap of the Africaner family for help, who came with some guns and stout fellows from near the Orange River, defeated the Damaras in three bloody fights, in 1835, took their cattle from them, conciliated the hill Damaras, and became the great chief of this part of the country.

The Namaquas were very civil. We felt ourselves quite at home among them, and were glad to see mat huts again. There was milk and honey beer in plenty; on which last Choubib, as was his wont, got very drunk and quarrelsome. I had some difficulty to keep my hands off the old fellow, for he insisted on my buying an ox from the man who had treated him, and who wanted some of our goods, which I did not wish to open till we got to Aramap's head-quarters.

On the 24th of May, Aramap's brother, with several other Namaquas, came on riding oxen, and in their best apparel to meet me, and to conduct me to the great chief. I left Naraes with them. We passed over one of the finest plains I had seen in Africa, covered with sweet grass, and with high trees, and bushes dispersed on it in detached groups, and among which wild horses were seen. We approached the banks of a river with a strange name, for such a scene, the Kei-kurup, or " First ugly river," and we found its banks rather steep, and with pools of water in its bed, which was about seventy yards broad. Looking across it, there appeared to be a great town of Namaqua and Hill Damara huts,

round and conical. The whole plain was covered
with huts, in hamlets of five and six together,
and cattle and sheep-kraals were beside there.
We had got then to "the fertile plains and fine
cattle country," which were laid down from native report on Arrowsmith's map, and I was much
rejoiced to think that the ship of war had not
come for us, or we should have missed seeing
the three hundred miles of new country we had
just passed over, after leaving Walvisch Bay,
and the very fine region for grass and game we
were now in.

The landscape, besides being beautiful from
the abundance of trees and pasture, (amongst
which large herds and flocks were seen grazing in
every direction,) was imposing by reason of the
picturesque and primitive mountains to the north
and east, and placing the town of Niais (or very
black) as it were in a vast amphitheatre. The
first mountain to the north had four summits,
and as it had no particular name, I dignified it
with that of the Hydrographer to the Admiralty,
as Captain Beaufort had been of the greatest
assistance to the expedition. The most distant

mountain, a blue peak, (Karubees, or Roll Mount,) was subsequently named at the Royal Geographical Society after myself, whilst south of this was Huhap, Thorn Glen Mountain, Hubies, (much, or) the Great Mountain, eighteen hundred feet high, square topped, with a peak at the southern extremity, and whose sides were deeply furrowed. South of Hubies was Nahabip, or Tortoise Mountain, and some minor heights. I named this group of mountains, the most picturesque I had seen, after our most gracious Sovereign Lady the Queen.

After enjoying the view of the detached mountains and of the plains at their feet, and calculating that in the scattered town of Niais, there must have been about one thousand two hundred souls, I crossed the Kei-kurup, and halted on the other bank, where I directed the people to unload the cattle. Aramap now came from his hut attended by several of his old people. He was a little, modest looking man, with the usual Namaqua features, as to high cheek bones, narrow eyes, and prominent lips, but his nose was slightly inclined to aquiline.

He had nothing in his outward man to denote the bold and intrepid warrior, who had beaten the formidable tribe of Kamaka Damap, and had thus saved the Namaquas of the Upper Fish river from annihilation. But Aramap, like other great commanders, though short, is distinguished by a daring mind, by good judgment, and by very active habits.

He said that it was unsafe to "pack off," near the river, for lions swept along it almost nightly, and had lately carried off both sheep and cattle from his people; accordingly we carried up the baggage, with assistance, to a clear space adjoining Aramap's hut, who erected mat screens to shelter the people, and who did all in his power to render us comfortable.

Here then was I now at Niais, far in the interior of Africa, but seated once more in my tent, and in the midst of abundance! It is true that we might be attacked by Kamaka Damaras, but having Aramap near me, who knew so well how to deal with them, I had no anxiety on this score. We might now have swam in milk if we had been so disposed; night and morning the

women brought us great quantities to exchange
for large eyed needles; Choubib also had oppor-
tunities for getting drunk on honey beer, and
though we had nothing in the shape of bread
or vegetables, yet of flesh we had plenty. Aramap gave me a handsome present of pack and
slaughter oxen, and of sheep. I gave him a
cloak, medal, pipe, shawls, axes, beads, handkerchiefs, &c. in return, and we became great
friends.

CHAPTER VII.

Inquiries about further Progress — Impediments—the Ka-maka Damaras and the Kallihari Desert—the Nosop River—Elephants—a beautiful Oasis—Queen Adelaide's Baths—a grand Reed Dance—Kilfuddie—the Kamaka Damaras described—their Names—Language — Compara-tive Table of South African Languages—Feats of Strength —a Damara Warrior—Hand to Hand Fighting—the Dress of the Women—The Huts—a Bed of Thorns—Religious Belief — Marriages—Burials—Manner of making Peace Manner of Trading with the Namaquas and the Portuguese —Provisions for the Road—Accident with a Rhinoceros —Birds of the Tropic—a grand Lion Hunt.

SEEING that I could now get supplied for our further progress, through the kindness of Ara-map, who abundantly proved himself to be as liberal as he was brave, one of the first inquiries I made was, " can we proceed further to the north or east; for I am extremely anxious to pass either through the Damaras of the plains, or to travel towards the Eastern Ocean ?"

" I am sorry that you cannot do either," said

Aramap, " or I might make an arrangement to accompany you."

" Cannot we send a message to the Kamaka Damap, to tell them we are coming as friends?"

" It is impossible, no one would take a message from this for any reward; for the messenger would be sure of death among a people so wild, and so exasperated against us for turning them out of this country. The Damaras cannot be approached from this side; they can only be seen by going by sea to the coast, and from thence communicating with them, though even then they might be suspicious of being carried off for slaves."

" Since you say it is so difficult to proceed from hence across the Swakop, what is to prevent us travelling towards the rising sun, there are no Damaras in that direction to interfere with us?"

" No Damaras," said Aramap, " but a desert without a drop of water for men or cattle."

" I know there is a desert to the north and west of Latakoo, which it is impossible to cross; but we are far to the north of the Bechuana country, and I was in hopes that we had got beyond the Kallihari desert also, and that a belt of well watered country stretched across Africa from where we now are, to the country of the Maquainas, and the settlements of the Portuguese on the east coast."

"This is not the case," replied Aramap; "none of the Namaquas or Hill Damaras have ever been able to travel to the east from where we now are, and no natives from the Bechuana country have ever come across as far up as where we are now. I have been at Latakoo myself, but to reach it from the banks of the Fish river, I was obliged to travel, first south, towards the Orange. I next went up it, and then by Griquatown, I reached the Bechuanas. In going east from this, after the mountains are passed, nothing but sand and bushes are seen; water, as I said before, there is none. One can get some distance to the south-east from this (though not as far as Latakoo), and find food and water. I went thus to the Nosop river (or river of the Bechuanas), a short time ago, twenty days south east of this, to shoot elephants, and I got thirty pair of tusks. I passed the Kusip and the Kuba-kop rivers on my way to the Nosop."

" What sort of a river is the Nosop?"

" It is a river with plenty of bushes and a stony bed. In this the elephants had made holes with their tusks, and into these they had inserted their trunks to drink. There are plenty of ele-phants about the Nosop, and it took two or three men to lift some of the tusks."

Many of my readers are aware that in Ceylon, where there is plenty of water and grass, most of the elephants are tuskless; whereas in South Africa, where many of the rivers are periodical, and water and grass are often scarce, almost all the elephants are provided with tusks for defence against the rhinoceros, to obtain water in the manner just described by Aramap, and to dig up the mimosas, to eat the sweet and nutricious roots.

Travelling to the north-east from Niais, the first day's journey brings one to the source of the Kusip river, and the second to Awaz, or the Bean Mountain, near which is the source of the Kuisip (from which river Niais is only one day's

journey distant), and at Awaz is one of the
sources of the Swakop. The Nosop also rises
here, and sweeping to the east, say the natives,
it takes a turn towards Latakoo, and then pursuing a southerly course, it empties itself into
the Orange river.

Three days from Niais, a remarkable and
beautiful spot is reached. A long green hill
terminates in a peak; from the north and south
sides of the hill descend streams from hot and
cold springs; those which issue from the southern
slope are lost in a plain of the richest verdure,
whilst the northern streams unite and form a
lake several miles long, with reedy margin, and
where many new birds and fish are to be found.
Another hot spring is to the west of the lake,
and from which issues the principal source of
the " elephant-abounding " Swakop.

The Kamaka Damaras were most anxious to
retain the springs about what I named Queen
Adelaide's Baths; and they fought there the last
battle with Aramap and his Namaquas; but
now no one occupies this beautiful oasis, the
constantly watered and evergreen plain round

the Baths. It is too much in advance for
Aramap, and without constructing high stone
kraals there, the Namaquas would soon lose
their cattle, and perhaps their own lives by the
Damaras.

The weather at Niais was delightfully cool;
the thermometer was at 65° at sunrise, and at 70°
in the middle of the day. "The rainy season,"
said Aramap, "is two months hence (in August)
in this part of South Africa."

Immediately after the arrival of the expedition at Niais, Aramap ordered a grand reed
dance to take place, and at least an hundred
women came before the tent, young and old.
A full band of reed-players blew and stamped,
as before described, the women clapped their
hands, sang, and ran round the players; and
there was dust and noise to our hearts' content.
There was one old woman here with ostrich
feathers in her hair, who was one of the most
persevering dancers I ever saw, for she danced
for two or three days after the above beginning.
There seemed to be no tiring her throat, palms, or
heels, and I thought at last she would meet with

the fate of the dame described in this elegant
Scotch rhyme—

" There was an auld wife, and they ca'ad her Kilfuddie,
And aa body said she wad gang to the wuddie,
But I think she deed in a better commaund,
For she danced her to dead at her ain house end ! "

Among the Namaquas of Niais there were
several Kamaka Damaras, of both sexes, pri-
soners of war, captives of the bow and spear.
The young men were square built, and the
finest specimens of bone and muscle I had
almost ever seen, whilst their skins shone like
polished ebony. The young women were tall
and graceful, and with features much hand-
somer than those of the Namaquas, or Hill
Damaras.

The men opened their mouths and showed me
the mark of their nation : it consisted in the loss
of the two lower front teeth, and in the two
upper being filed, so as to leave an opening
between them like the letter A.

I questioned these Kamaka Damaras till I
tired them, regarding their people beyond the
Swakop.

The Damaras of the plains are a great nation, and their country is full of cattle—"which," say the Namaquas, "they get from a cave as they require them; so there is no great harm in our taking a few from them now and then, as they can easily make up the loss." They call themselves Oketenba Kacheheque, or Omotorontorondoo, and their language is soft and pleasant to the ear, abounds in vowels, and is quite free from clicks. I subjoin a short table of it, also of the Namaqua language, of that of the Amakosa Caffers, and of that of Delagoa Bay, the first time that these four have been compared.

TABLE OF DIALECTS.

	Damara.	Namaqua.	Amakosa.	Tembé.
A man	omdenkoonee	kooep	indota	moofo
A woman	inkonum-omkodentu	taras	umfas	umfat
A child	omaché	'toaee	umtooana	munwhana
Father	otalé	saöp	bao	baba
Mother	komama	saoos	maa	mama
Brother	omtenaquainke	sa'kap	nenawé	munaguam
Sister	omtena	sa'kas	dadewabo	tatooam
Ground	esi	'thoop	goomthlaba	moothlaba
Grass	ethodoo	'tkap	ee'na	booeané
Tree	omobontee	heip	goomtee	mootẻ
Water	omeba	'kums	goomansee	gumatee
Fire	omoreero	'eis	umdeloo	moolelo
Flesh	oniama	'kunee	iama	inyama
Fish	oseema	'oup	intlansee	inthlante
Fowl	ontera	aneep	intaka	inyonee
An ox	ontoombee	koomap	inkabee	eeabee
A cow	onkobententoo	koomas	imasee	eeomogätee
A bull	ontoentoo	'kop	inconsee	eeontee
A calf	ontana	champ	itolee	eeunyanä
A sheep	ondontoo	koos	goosha	eenfee
A lamb	ontona	korooee	inkoeana-igoosha	eentenyanee
A goat	ononkonbo	poorees	bok	eembootee
A kid	okankombo	pooreeroee	inkoeana ibok	letenganee leembootee
A horse	onkora	'häp	yehash	*
Milk	omeisi	deiee	amas	loobẻs
A dog	omva	areep	eenja	eenja
An elephant	ontòoo	'koap	inglovoo	inthlov
A rhinoceros	onkaba	'nabas	oomkonbee	shoo sho shoo
A snake	onoka	'aop	eeoka	inyoka
A horn	obona	'naop	impondo	loopontoo
Tobacco	kumakaéa	'tubaccaee	eekooba	efölee
An assegae	enka	'kooap	goomkontoo	moó ónoo
A kaross	oombanta	'numee	ingooboo	inkoobo
Good	omdehenta	esahooee	nobom	koothley
Bad	ochunkoontee	'hoatoma	embee	goobee
Big	omodee	kei	inkooloo	ookooloo
Little	okateeché	'kureea	'inkanee	goonanee
Warm	omebaté	'kumchee	eshooshoo	peeachesa
Cold	ombepera	'keia	yebanda	komo kato
One	emoonjé	'kooé	senyé	eenya
Two	embaré	'täm	subené	impeytee
Three	datoo	'oona	sitatoo	inatoo
Four	ené	haka	zené	moonee
Five	indan	koree	silthlanoo	ithlan
Six	eambomé	'nanee	siput thlongo	tegatoopakoo
Seven	eambombaree	hoo	isikéiné	natumbetoo
Eight	eambomdatoo	'keisa	siput thongo	toopakool natim atoo
Nine	eamboené	koësé	sitoba	nenthlandee
Ten	himorong	deesee	ishoomee	eshoomo.

From the above specimens it will be seen how little similarity there is between the Damara and Namaqua languages, and again how much they differ from the Amakosa and Tembé languages, which last are different dialects of the same language. The inverted commas are the clicks of the tongue.

As every thing connected with the Kamaka Damaras is exceedingly interesting, I beg to subjoin a few more words of their language.

Sun	eiooba	Hard	obekokooto
Moon	omedé	Soft	obetarahoo
Wind	obebo	Sweet	ochewa
Cloud	oboora	Sour	obĕrooroo
Head	onacheesce	Dry	quakaha
Arm	okoó óko	Wet	okokoturet
Hand	omaké	High	okokoorce
Eye	omeho	Low	obesh
Ear	homachooe	Sharp	okohoka
Tooth	omaéo	Blunt	ojeboomo
Blood	obetoo	Black	oche norondoo
Day	herero	White	oche sorondoo
Night	oudook	Red	oche serondoo
War	obet	Yellow	ombaka
Peace	koonahangé	Blue	oche honee
Weary	bou ogwa	Green	stadoo
Hungry	tatochera	Angry	owapindeke
Thirsty	patenyouda	Sick	weiberee

The Kamaka Damaras are believed to be the most muscular nation in South Africa, from the healthy climate they inhabit, and from their good living, for they have always plenty of flesh. I could not ascertain that they cultivated any thing except tobacco.

The stories told of the feats of strength of the men are strange enough. Thus, when they wish to kill an ox, one man, it is said, will take it by the tail and swing it round, and thus throw it down, or he will take it by the horns and throw it on its side, then turning it on its back, and placing his knee on its breast, and grasping its throat with his hands, he strangles it. This way of killing cattle may probably arise from a desire to save the blood. In another work, I described the extraordinary manner in which the Caffers kill an ox, that the blood may not be wasted.

When a lion kills any thing belonging to a Damara, the people of the village assemble, and go out with the man who has lost his property; but he only must kill the lion: the others are merely spectators.

The *naked* Damaras is another name for the Kamaka Damap, for when not engaged in war they only wear fore and after aprons of softened skin. In war their appearance is picturesque and formidable — a bunch of ostrich feathers waves on their brows, their woolly hair is worn long, over their shoulders is thrown a kaross of lion or leopard's skin, a leather band is twisted many times round the waist to support them, and in this is stuck some arrows when they have begun fighting, and have thrown away their skin quiver; in their right hand is a short and heavy club; (one which I have got is of dark red wood, and looks as if it had been turned), in the left hand is a long bow, also a strange broad-bladed javelin, five feet long, the shaft and blade composed of iron, and without any wood work about it, (half of the shaft of one in my possession is cased in leather, that it may be grasped more firmly), sharp pointed sandals are on the feet, ivory rings and beads are on the wrists and neck.

After a discharge of arrows the Kamaka Damaras close with their enemies, attempt to knock out the brains with the club, and stab upwards with their iron assegae, which is not intended for throwing. The Zoolas of the country behind Natal also use one assegae (with a *wooden* shaft), and make a desperate onset with it, like Highlanders with broad swords, whereas other South African nations hurl their light assegaes from a distance, and therefore are not half so formidable as the hand-to-hand-warriors.

" Like waves of ocean rolling fast,
    Or thunder clouds before the blast,
    The swarthy legions stern and vast
    Rush to the dreadful revelry."

The women of the Kamaka Damaras wear a conical cap of softened skin, with loaf-shaped pieces of stiff leather all round it. They also have got a fore flap of skin depending from the middle, and a hind one; the former is the longest. They wear copper ornaments, and rows of ostrich shell beads arranged *perpendicularly* round the waist.

The huts of the Kamaka Damap are small and conical, are composed of stakes wattled with branches, and plastered with mud, and are covered with hides. The entrance is low, and the huts are comfortable and warm, owing to the want of karosses among the inmates. The bed is composed of branches of thorny mimosa, on which skins are spread. This strange couch may have been adopted to guard against snakes. The

Kamaka Damaras do not cook in their huts, but outside, and the men do not eat with the women.

The Kamaka Damaras believe in a Great Spirit, in one powerful unseen being, of whom they are afraid.

When a couple have determined on marriage they go together to the young woman's father's house, when he kills an ox or more to feast on; the party then goes to the bridegroom's place, and there he kills as many head as his father-in-law did, and thus the marriage is concluded.

When a man dies two oxen are killed, of the flesh of one of them the young men only eat, the other is an offering for the dead; and in the hide of this last the body is buried, in a sitting position, whilst the ox's head and horns are set in a tree over the grave.

When the Namaquas wished to be at peace with the Kamaka Damaras, they went with unpeeled sticks in their hands, and with the feathers of a spotted duck in their hair, and sat down on a hill opposite a Damara village. Their late enemies then came to them, and taking the

Namaquas by the hand, they led them to water, which they threw violently over them, perhaps to wash away all enmity between them. The Namaquas were then taken to the village and distributed in different houses, but were not allowed to wander about spying, though the Damaras assumed that privilege when they came among the Namaquas.

Each Damara host then killed a beast for his guest, and it was expected that it should be all eaten up, (which the Namaquas were, doubtless, never loath to attempt to accomplish); and when the Namaquas departed, each was presented with a lean calf, which he was to rear carefully, and it was understood, that as long as it lived, so long was friendship to continue between the donor and the receiver of the gift.

In trading with the Damaras, and exchanging an axe for an ox, for instance, and the ox is killed for a feast, the Damara will look for half a day at his purchase, and if at last he discovers a flaw in it, he will immediately bring it back, throw it down, and take up what remains of the ox without saying a word.

" We were in the habit," said one of the women, (a fine negress), " of going to an inlet of the sea, where the women left the men with the cattle, and were taken across in a boat to the other side, where white men wearing hats were seen; these people we call Oban in the Damara language, another name for ourselves is Obantoo-bororontoo, and the Namaquas we call Obatchna."

" With the Oban we exchanged our cattle for iron to make javelins, for copper to make beads, and we also got knives and calabashes from them. We would not allow the Obans to come into our country. If the Damaras go to white people they go as friends; but the Damaras will not allow white men to enter their country, if they do, they must be considered as enemies."

Where the Creek or Bay is, to which the Damaras resort to trade with the Portuguese, it would be interesting to find out, for at that point they could best be communicated with, and as they are a fine manly race, and as I heard of no bloody rites or cruel practices among them, a stranger might be safe among them, if he

managed in the first instance to disarm their
suspicions. Great Fish Bay is in 17° south
latitude nearly, possibly it may be there, where
the Damaras see the Oban.

As we had still much ground to go over, which
ever way we steered, I was resolved that we
should not again be on the verge of starvation,
as we had hitherto often been, accordingly I increased my flock to the number of 160, and for
each new sheep or goat which I purchased, I
paid according to Aramap's desire and the
country practice, a knife, a handkerchief, or a
tinder-box, that is less than sixpence a-head. I
also bought some oxen for gown pieces, shawls,
axes, beads, &c., as I anticipated that some of
my old and worn-out cattle would soon perish by
the way.

It was a fine sight in the evening, the herds
and flocks returning from their pastures, where
they had been grazing in company with zebras
and steenboks, (by creeping amongst a herd of
cattle one day, Elliot shot a male zebra), clouds
of dust rose on every side on the plain at sundown, and cattle and sheep in thousands were

seen pouring towards the huts, that they might be safer from lions there, than at a distance from the men and dogs.

Some of the people went after a white rhinoceros one day, not far from Niais. They rapidly pursued it, and it fled before them, when, in passing a clump of bushes, a black rhinoceros rushed out on the hunters; they scattered themselves immediately, and attempted to escape, but the last of the party was caught on the terrible horn and thrown into the air. When he fell on the ground on his back, he had the presence of mind to lie quite still, and with his eyes nearly closed. The brute then made short rushes at him, snorting and smelling to ascertain if he still lived, but he lay as if dead, for he knew that if he had moved, he would have been instantly gored and trampled to death. In the meantime the rest of the hunters, missing their companion, turned, and seeing what the rhinoceros was about, they gave him a shot in the breech, which sent him off screaming, and with his tail between his legs. The man who had been tossed was brought to me, but the only injury I found he had sustained was a slight graze on one leg, and which a simple plaster put to rights.

We shot many beautiful birds here, particularly coracias (with crimson and white speckled breasts, and blue wings), a short tailed species of which, as I before mentioned, alights on the horn of the rhinoceros. The birds were here more brilliant in plumage than we had yet seen, and doubtless the feathered tribes would have gone on increasing in beauty and in variety, if we could have got further beyond the tropic, and approached nearer the line. I was glad we had got so far as we had done, and to feel myself fresh and alive. No one can accomplish all he attempts in this world of trials.

It was on the evening of Sunday the 28th May, that I was pacing in front of my tent, and enjoying the starlight—whilst the people were wrapping themselves in their sheep-skins and lying down to sleep, contented and happy, after plentiful meals of flesh and milk, and a social pipe—when I heard a woman's voice calling out at some distance in the Namaqua language, " Keiré huéteyré!" (stand up and help, friends!) then a great fire was made, and the dogs began barking loudly. I roused Henrick Buys, and we tried to find out what was the matter, but could not. The cattle seemed uneasy, and moved about; but in a little time all was still. "It must have been lions," said Henrick.

Next morning we found the traces of three lions coming straight to the tent, but they had been turned off by the dogs. They had then made attempts to get into the town at two other places, but had been again scared by fires, and the faithful guardians of the night.

At eight o'clock a man came in from the field, and reported that one of my oxen, Zwart Berg, the oldest of the herd, lay dead on the river's bank, and in part devoured. Berg had strayed the day before, and could not be found to be driven in with the rest, and therefore had fallen a prey to the lions.

I went to Aramap, and asked him if he was inclined for a lion hunt, and he said that he was quite ready if I would furnish the ammunition; accordingly, a strong party was mustered, twenty

muskets were loaded, and those unprovided with
fire-arms, were equipped with lances, bows,
arrows, and clubs.

We proceeded to the spot, above which crows
were wheeling in the air, and indicating the feast
they had in contemplation below. Under an
aged tree lay the bullock on its side, with marks
of teeth on its throat, the stomach was also torn
out, and the skin and flesh stripped off the hind
quarters. We looked about for spoor, and found
that two of the lions had gone south and another
east; we followed on the traces of the two, like
hounds tracking game by the scent.

The people went over the ground at a rapid
pace, talking and making some noise, when,
about a mile from the dead bullock, a large male
lion started out of a bush, and galloped off before
us, among the trees and underwood. A shout
was immediately raised, by the whole body of
hunters, who with the dogs unhesitatingly gave
chase.

We ran down hills, along valleys, and up
acclivities, leapt rocks, and brushed by thorn
bushes. Sometimes we lost sight of the lion;

but again he appeared galloping and trotting
on before us, with the dogs close at his heels.
We ran for an hour, and our strength was tried
to the uttermost with the broken ground.   I was
rather inconvenienced by some extra ammunition I carried for the people, and I thought the
lion was never to be brought to bay, when he
began to slacken his pace.   Some of the people
of Aramap then ran ahead, cut off an angle, and
the lion seeing people before him, "brought up"
in a large bush.

The dogs now surrounded the bush, and
barked in a tone of mingled fear and anger, but
dared not go in, for the monarch of the plains
raising himself, and with mouth open, mane
erected, and his tail lashing the ground like the
flail of a thresher, growled and roared terrifically.
I had previously told Aramap that only half the
people must fire at once, and the other half
reserve their fire; and the line was in the act of
being hastily formed for a volley, when the lion
again dashed off at right angles to his former
course, with all the dogs after him.

The hunters had recovered their breath for

another run, when a noble black dog, foremost in the pursuit, bit the heel of the lion. Before this, the king of beasts, flying before his pursuers, looked like a large yellow mastiff; now he altered in a moment, and with every hair on end, with teeth displayed, and tail in the air, he wheeled round and swelled out apparently to the size of an ox. Despising the dogs, his whole attention was bent on the men—he rushes towards them from a distance of two or three hundred yards—there is a cry of *'gnoo! 'gnoo!* (sit! sit!) for it is of no use standing up against a lion or running away from him—almost all are down in a moment, for a lion is less likely to strike a person standing up than one sitting or lying, and if he does strike, he remains on his victim, so that the rest have then an opportunity of destroying him. We made up our minds to one of the party being sacrificed.

The lion dashes through bushes, nothing stops him or turns him aside, he springs into the air over a rock, and tears off a portion of it with his hind claws—the word is given to fire, and a ball takes effect in his left shoulder—he rolls over twelve yards off, but immediately recovering himself, he again comes on frantic with rage and pain, another volley is given him, and he falls dead at the distance of four yards only from the muzzles of the guns, with balls through his head and side.

The hunters now breathe more freely, crowd round the lion for a moment, and then, leaving him, they begin to recount their manner of dealing with him, or to laugh at the fears of some of the party. My whole attention was directed to save his skin, and beating off the dogs (who were beginning to pull his mane) with the butt-end of my piece, and, assisted by Robert and Antonio, I soon had his hide off. The Hill Damaras then cut up his carcase, and carried him off joyfully for their evening meal. I tasted him also, and he was rank and unpleasant. Two men carried with difficulty his head and skin on a stick to Niais, eight miles distant. He measured ten feet from his muzzle to the " tossle" of his tail.

VICTORIA MOUNTAINS.

Etched by William Heath —

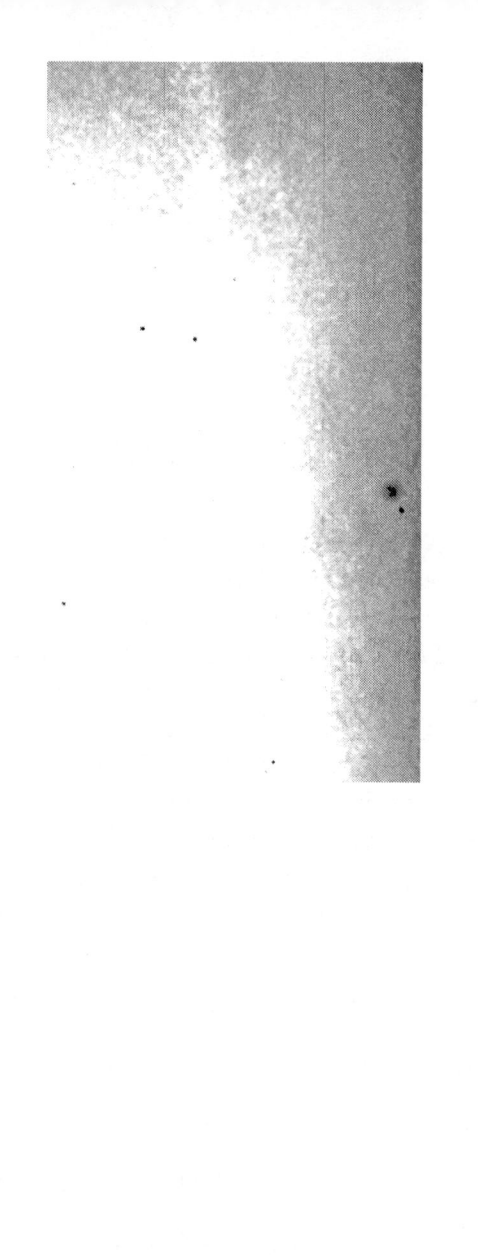

CHAPTER VIII.

The Damaras of Niais—Disunion—The Pot Dance—African
Travelling—Leave Niais—The fine Valley of the Bath—
Good Situation for a Mission Station—A fearful Tragedy
—Koodoos—The Copper Mine of Gnutuais—A strange
Region—Leave Damara Land—Robert makes a narrow
Escape—The Calf River—A Murderer and a Witch in
our Company—A Tale of Witchcraft—the Water Bull—
Lions—The Gum Pauw—Cameleopards—A Disappoint-
ment—Native Manner of hunting the Giraffe—The Lion
and the Cameleopard—The Great Fish River—Kuis—A
Fishing Party—New Fishes—Notes for Naturalists.

AT Niais, the Hill Damaras lived apart from
the Namaquas. I visited the former, and found
that they had got the charge of cattle and sheep
belonging to the latter, but they seemed here as
ignorant as I had before found them. I brought
their chief to the tent to question him; he
seemed to have been chosen for his size and
strength only, as to mind he seemed to have
none. In talking to him about the generally
impressive subject of death, he merely said,
" If I live, I live; if I die, I die; but the lambs
are in charge of the children, I must go away

and look after them;" and then held out his hand for tobacco.   .

One afternoon I heard angry words between Old Choubib and Magasee; they had quarrelled about a pipe of tobacco. Choubib, with his usual greediness, not content with his own allowance, wanted Magasee to give him a pipe full of his also. From words they were proceeding to blows, that is, they were stripping to wrestle and to give one another as heavy falls as they could, which might have brought on a general battle between my Cape people and the Namaquas, the consequences of which might have been fatal. Being fortunately near, I prevented the angry wrestling.

The people had been agreeing among themselves pretty well of late, still it was not possible always to make them act up to this maxim, that as our undertaking was one so should our hearts be likewise.

Cor unum, via una.

As our arrival at Niais commenced with a dance, so our sojourn there also ended with one. The pot-dance, which I had not yet seen, was performed. About thirty Namaqua women seated themselves in a hut, from the arched roof of which hung two chords; these were grasped by a man, who commenced stamping the ground first with one foot, and when that was tired, changing it for the other. He also sung in low chorus, " Uwahu," to the " Ei, oh; ei, oh! ei, oh! ei, oh! ei, oh—oh! oh! oh!" and clapping of the hands of the women. One of these held before her a bambus, in which was a little water, and over the top of it was stretched a piece of sheep-skin. This was occasionally wetted with the water inside, and was beaten with the fore-finger of the right-hand, whilst the pitch was regulated by the fore-finger and thumb of the left.

The dancer tried occasionally to slip off the skin head-coverings of the women, and after he had danced his fill, his place was supplied by another. It was pleasing to see people so happy as these were, and so innocently engaged. There was not the least impropriety observable in this dance; it was merely harmless excitement, and abandonment to the mirth inspired by the most simple of all music.

The night before we left Niais, a horn was blown to arouse the sleepers, and a rocket was let off " to astonish the natives."

On the 31st May, the cattle and sheep being collected, and the packs in order, we left Niais, my people now joyfully turning their heads to the south. I had set out originally from the Cape with the intention of getting as far to the north and east as I could, without laying down for myself any particular parallel of latitude or meridian of longitude to reach. It is less possible to say, in African travelling, than in that of any other country, that we shall reach such and such a point before we turn. Difficulties and dangers everywhere abound in the interior of Africa, and he is very lucky who can traverse a considerable portion of *unmapped* regions, and return to tell that he has done so. I would have gone to the line if I could, or to the Indian ocean from the Atlantic, but it was impossible for me to do so, as I have already sufficiently explained. I hope, therefore, I shall not be blamed, if, by not " running my head against a wall," and having my people destroyed beyond the tropic, I am now enabled to tell what I did see, and what I did accomplish. I did my utmost to make an extensive sweep—man could do no more. I have no doubt, that my own seven men would have gone to the north with me, and also Henrick Buys, who said he would follow me to the death anywhere, but none of the Namaquas, or Hill Damaras, who knew the country and the waters, would, as I before said, go a step towards the Kamaka Damaras.

The greater part of the population of Niais turned out to see us depart; and the people shouted to cheer us on our way, whilst a great many of their woolly heads were ornamented with our gaudy cotton handkerchiefs. The excellent Aramap, his fine large wife, and several of his head people accompanied me, to point out a copper mine on our road. We went down the Keikurup river for twelve miles, and then outspanned on its banks at a hamlet.

On the 1st June we had a long march, but it

was through a very beautiful country, abounding in trees and grass. The remarkable Bid Stone and White Mountains, were passed on our right. We descended a ridge, and found ourselves, after twenty-six miles walking and riding, and twenty-four hours without food, in a very fine valley, stretching east and west for several miles, and two or three in breadth. It was enclosed with hills on the north, south, and west; towards the east it was open, and in that direction it afforded a distant prospect of some steep and lofty mountains,* terminating in sharp ridges.

There was quite a forest of thorn trees of several miles in extent in the valley, and in which also the grass stood like corn. Pheasants and Guinea fowl ran about amongst the grass, in large flocks; deer were among the stems of the trees, and tuneful birds were on the branches, so that I could not help repeating—

* One of which I named after Sir George Grey, Bart., Under Secretary of State; and another after Robert W. Hay, Esq., F. R. S., late Under Secretary of State for the Colonies.

"'Tis merry, 'tis merry, in good green wood,
   When the mavis and merle are singing,
When the deer sweeps by, and the hounds are in cry,
   And the hunters horn is ringing."

In the centre of the valley among rocks of granite, rose a warm spring of 126° of Fahrenheit. The water flowed freely from it, and was lost in the plain. Though there were no people in this happy valley, to take advantage of the water, yet Aramap said it was a favourite resort of his, and he shewed me where he had made a dam, and had cultivated some calabashes and tobacco. I gave him some melon and cucumber seeds, to extend his garden. Maize might be raized to a considerable extent here, by leading out the water of the Bath, which, as it had no designation, I named after Lord Glenelg, her Majesty's Principal Secretary of State for the Colonies.

I was so delighted with the valley of the Bath, abounding as it did in water, grass, and trees, and with a striking view toward the east, that I remained a couple of days in it. Two new birds, afterwards described, were here

added to our collection, and a white rhinoceros was shot in the valley; but a party of
Hill Damaras, on the look out, carried off
almost all the flesh before the pack-oxen could
be sent from the Bath to the place where the
rhinoceros fell.

As the valley is such as I have described it,
capable of supplying abundantly the necessaries
of life, being also in the midst of game, and of a
numerous population of Namaquas and Hill Damaras, it seemed well adapted for the establishment of a mission station, and accordingly I have
indicated its existence and its advantages to the
Wesleyan Missionary Society, as, although a distant point, it might yet be connected with their
present stations at the Kamiesberg and Nisbett's
Bath, or a more direct rout than that which
we had lately traversed, might be found from
Glenelg's Bath, to a station which may ultimately be formed at Walvisch Bay.

There is one thing which the friends of
missions ought to be aware of, Aramap and his
people are anxious for missionaries, the women

in particular said, " send us teachers for our-selves and for our children," we ought therefore speedily to respond to their appeal.

We saw indications of the Kamaka Damaras at the New Bath, such as the head of an ox placed on a tree to mark the grave of a warrior below. Here, too, some time ago a fearful tragedy had been enacted. The Namaquas living at the Bath, were aroused one night by the howling, as it seemed, of wolves, when imme-diately after, there was a fearful rush of negro savages, who destroyed the women and children, whilst the men, less encumbered, escaped to a neighbouring hill. The wife of the head man was secured by the Damaras, and had her hands cut off by the wrists next morning, before her own cattle, previous to her being put to death.

Under the protection of Aramap, a missionary would have little to fear from enemies of any kind.

On the 3d of June we crossed the Tell Tale river, which is lost in the sands to the east, and in passing over some undulating ground several of that most magnificent antelope, the Koodoo, were observed. A noble male came bounding toward the hunters, who ran toward bushes and lay down to surprise him—he tossed his mighty horns in the air, and wheeled to retreat, as he saw a head rise near him, and heard the cocking of a rifle; but he was too late, a single well-directed ball took him in the shoulder, and he fell heavily to the earth.

The koodoo is a large bodied buck, and the limbs are strong in proportion to its bulk; its height is four feet, its general colour is light brown, with a narrow white band down the back, and about eight white stripes proceeding from that and descending on the sides; the koodoo is provided with a mane, and its tail is of some length; but its glory is in its horns, which are superb. They are commonly four feet long, and spread out from the roots in two or three spiral turns; their colour is brown with white tips. The koodoo is always near cover, and is seldom or never seen in the open plains.

Twelve miles brought us to the Konap or Dry river, and to a place in it called Gnutuais or Black Mud, where we found small pools of water. A mile above these, on an eminence, Aramap pointed out the copper mine, of which we were in search. There was a long trench in the ground, in which the Damaras had been digging for metal to forge rings between two stones. Sandstone and quartz were about the trench, and every stone was covered with verde-grease.

There is undoubtedly a good store of copper at Gnutuais, near the surface of the ground, and there is also wood for smelting it, and water for the supply of the miners; but its distance from the colony and the sea, may prevent it being made available for some time to come, at least not till the copper near the Orange river has had a fair trial. If the Namaquas had a blast bellows they would soon run copper balls at Gnutuais.

I now parted from Aramap with considerable regret, for he had proved himself to be very intelligent and civil, and was really most hospitable. To himself and his wife I gave, as parting gifts, the best of what I could spare, as saws, axes, garnet beads, gown pieces, &c., and I hope they both left me contented.

Continuing our journey, we passed through a strange country, full of rocks and trees—the rocks no where rose into hills, but were merely immense boulders lying on the surface, and in some cases were piled on each other. We crossed the Kubieb or Stick-grass river, the Huerap or Crooked river, and offpacked, after four and a half hours, at the Tuap or Clay river, near which two more rhinoceroses were destroyed by a party of hunters travelling the same way as ourselves. Some detached rocks at the Tuap stood like pillars in the wilderness, and the general appearance of the landscape was very pleasing.

On the 5th, the country we traversed was a vast plain, with hardly a rise to be seen in any direction, but everywhere there was most abundant pasture and water in pools. Passing the Nonowus or Hornback hill on our left, and seeing the traces of large game, we offpacked at the Ku Kama or Brown river. Here on many trees swung the nests of ingenious weaver birds, safe from baboons and snakes, at the extremities of the branches.

We now left the Hill Damaras all behind us, though they extend farther to the south, more to the westward.

We again entered *veritable* Namaqua land, with patches of sand, quartz, dry white grass, and bushes. Robert now made a narrow escape. Being among the last of the party, and seeing a rhinoceros on the plain below him, he called out to some of the Namaquas near him to follow and help him to kill the animal. But they, seeing that there was not sufficient cover for them to " be-creep" the rhinoceros with safety, allowed Robert to go on alone. He did so on foot; and as the rhinoceros was on the move, he followed it for some time till it went amongst some thorn bushes. Robert approached them, when the rhinoceros, again appearing, saw Robert for the first time, and immediately gave chase. Robert ran off as fast as he could; the rhinoceros fol-lowed, snorting and tearing up the ground with its horn. Robert threw off his hat to try and stop the brute; but it disregarded it, and after a run of a couple of hundred yards, a bush luckily came in the way, into which Robert threw himself and lay flat. The rhinoceros passed close to him, and missing him, stopped to look round. There was no time to be lost; Robert fired and struck the rhinoceros in the neck; it fell, and struggled to get up. Robert took to his heels again, picked up his hat, and joined me in advance quite out of breath. If he had allowed the rhinoceros to recover from the stunning effects of the ball, he must have been destroyed.

A troop of koodoos again bounded across the slope before us; a male with six or seven hornless females were seen. The leader was selected by the hunters, a brace of balls crushed through his side, which, with a charge of buck shot in his muscular neck, caused him to spring into the air, and to fall panting and sighing on the ground. Though we had now sufficient sheep for our support, it would have been overstrained humanity not to have " shot down " a fat buck when it came in one's way, to vary our diet, and to keep the people exhilarated with " sport."

After thirty miles we packed off at the Chounp, or calf river.

We had now in our company an ugly little old Boschman—wirey and wizened, with the broadest nose, the widest mouth, and the deepest wrinkles about the eyes, I ever saw. Henrick Buys called him "Hornkop," or horn head, from a strange bump or excrescence on one side of his head. His wife and child were also with us, and were as little attractive as himself. This amiable family had asked leave to travel with us towards the Fish River, whither they were bound.

Hornkop lay all night near the dead koodoo, and a lion came and ate off its backbone, and when the Boschman appeared alive in the morning, the Namaquas said he was too great a rascal for the lion to touch him—it disdained to dirty its mouth with him.

I asked how he got the bump on his head.

"He is a murderer," said Henrick, "and he got the mark of a beast in this way. He had a quarrel with a nephew of his, and he continued his spite against him for a long time, till an opportunity offered for him to be revenged.

At last, the two were on a honey hunt together, and the nephew went into a hole after the bees, leaving his bow and arrows outside. Hornkop immediately fixed a barb, and shot his nephew as he came out, and killed him; then bruising his own head with a stone, he went and told his people that his nephew had done it, and that he had been obliged to kill him in self defence."

" His wife, too, looks like a devil," I said.

" Yes," said Henrick, " She is a witch, and can turn herself into a wolf if she likes."

" Do you believe that?"

" It is believed in the land that some of the Bosch-people can change themselves into wolves or lions when they like. They say of Hornkop's wife, that she was jealous of another old woman some time ago, and one evening, when the old woman went for water after dark, she never came back again to her hut. In the morning the people went out to look for her, and they found her bones beside a bush, and the footmarks of a wolf all about, whilst her skin clothes had been rolled up and placed at one side, not in the way that a common wolf would do, but in a way that Hornkop's wife must have done."

Here is another story of witchcraft:

Once on a time a certain Namaqua was travelling in company with a Boschwoman carrying a child on her back. They had proceeded some distance on their journey, when a troop of wild horses appeared, and the man said to the woman, "I am hungry; and as I know you can turn yourself into a lion, do so now, and catch us a wild horse, that we may eat."

The woman answered, "You'll be afraid."

"No, no," said the man; "I am afraid of dying of hunger, but not of you."

Whilst he was yet speaking, hair began to appear at the back of the woman's neck, her nails began to assume the appearance of claws, and her features altered. She set down the child.

The man alarmed at the change, climbed a tree close by, the woman glared at him fearfully, and going to one side she threw off her skin petticoat, when a perfect lion rushed out into the plain; it bounded and crept among the bushes towards the wild horses, and springing on one of them, it fell, and the lion lapped its blood. The lion then came back to where the child was crying, and the man called from the tree, " enough ! enough !—don't hurt me. Put off your lion's shape, I'll never ask to see this again."

The lion looked at him and growled. " I'll remain here till I die," said the man, "if you don't become a woman again." The mane and tail then began to disappear, the lion went towards the bush where the skin petticoat lay; it was slipped on, and the woman in her proper shape took up the child. The man descended, partook of the horse's flesh, but never again asked the woman to catch game for him.

The Namaquas believe that in certain deep fountains there lives a water bull, which is black and has large horns; at night it comes out to eat grass, and dives under the water before day. There is a similar superstition in Scotland.

We found the Calf river a disagreeable place, but we were forced to halt here for two nights to rest the cattle. A black hawk was shot here

by Elliot, the only one we had seen on the
journey. After dark, on the second night, two
lions growled round the trees where we lay.
The sheep were restless, and on rising to turn
them in near a fire, old Aaron, the *soi disant*
guide, with the invincible appetite, was discovered devouring the flesh of a jackal, which
had been shot during the day, and with jaws
chattering from fear of the lions, he said he did
not see why good flesh should be wasted, and
that he should sit up and do nothing.

On the 8th, our course was S.S.E. along the
Chounp, we crossed it, and off-packed at a
pool called Hoakosams, or scratch-breast place,
twenty miles. Here a gum pauw, or bustard,
which subsists partly on gum, was shot, which
measured eight feet and a half between the extremities of the wings, and was four feet seven
inches from the point of the beak to the end of
the tail. With its long legs it must have stood
upwards of four feet high. Its back was of a
brownish colour, its crest black, its neck speckled
with grey and black, its breast and belly were
while, and the shoulders of its wings were beau-

tifully marked with black and white feathers. This bustard was the largest bird we had seen since we left the Cape, and as everything " was fish which came to our net," we ate it, and found it very good.

We had not been long halted, when the word was passed that two cameleopards were in sight, and on looking towards the top of a ridge about a mile from us, two of these most graceful and long necked animals, were seen gazing towards us, and "craning" over the tops of the bushes.

The flesh of the giraffe is very excellent eating, accordingly we quickly seized our arms, and ran towards the ridge, concealing ourselves as much as we could. The main body of the hunters, doffing their clothes and shoes, stole along the bottom of a ravine.

We got near the cameleopards before they disappeared, and on attaining the summit of the ridge, I saw the two videttes who had been out reconnoitering, galloping back to join the main body.

It was one of the finest sights which I had ever beheld, to observe twelve of these most remarkable animals, small and great, on an open part of the bushy plain below me; there they stood in various attitudes, as if consulting how they could best escape their enemies whose vicinity they had just become aware of. They seemed to hesitate, and one or two lowered their heads to snatch a mouthful of grass; the hearts of the hunters beat high with anxiety, and they had got almost within range, when Elliot, who had never seen cameleopards before, and thinking he could run up to them, in his eagerness to get a shot, dashed across an open space, when instantly the troop came together at a walk in their strange manner, by moving two legs on the same side at once (different from other quadrupeds, which move opposite legs simultaneously in walking) they then broke into a gallop, and went off sawing the air with their long necks, and carried their brown spotted bodies far beyond our reach.

The Namaquas were much annoyed at losing a giraffe feast which they made sure of, and of getting some capital sole leather; but we afterwards ate cameleopard flesh, which we found

K 3

more juicy and palatable than any of the deer tribe we had partaken of in South Africa. ·

The most approved manner in the present day of hunting the giraffe in Namaqua land, is on horseback. The natives watch the giraffes when they are reposing, and get as near to them as they can before they dash forward; the gigantic beasts get up in their scrambling manner, and move off as fast as they can; the horsemen pursue them, and if their horses are good, they are not very long in coming up with them, and they fire when the muzzle almost touches the shoulder of the giraffe. When the cameleopard is mortally wounded, the long neck begins gradually to droop, and then the inoffensive animal falls on the ground. Sometimes, however, an old giraffe will stop short in his flight, allow the horseman to pass close to him, and rearing up, will overwhelm both horse and rider, and then make off.

The death of the cameleopard by a lion must be a remarkable sight. The lion is said to lie in wait for his giant prey among the reeds at a fountain, and when the giraffe stoops to drink, his enemy springs on him, when the affrighted animal carries off his terrible burden over the plain, the lion biting him all the while till he sinks and dies—

> " Plunging oft with frantic bound,
>   To shake the tyrant to the ground,
>   He shrieks—he rushes through the waste
>   With glaring eye, and headlong haste—
>   In vain!   The spoiler on his prize
>   Rides proudly—tearing as he flies."

In the evening, before going to sleep, we had set fire to an old tree near our lair, to keep off the lions; it fell in the middle of the night with a terrible crash, and the frightened bullocks ran over us. Some of the party were for firing, thinking we were attacked by a troop of lions, and it was fortunate that no one was hurt, either by the fall of the tree, the feet of the bullocks, or by an indiscriminate discharge of fire-arms.

Next day, after nineteen miles journey, we got once more to the 'Oup, or Fish river, and to a place called Hatep, or Reed Water, where we packed off under some remarkably bent trees. I went with Robert in the evening to fish and shoot at a large pool lower down the river, and at sun down we had a sharp walk home, expecting to meet a lion in the path; but we met nothing but guinea fowl going to roost.

Eight miles E. S. E. along the Fish river, brought us to Kuis, or Scent, the place of a Bastaard of the name of Kraai. We found about a dozen huts at Kuis, and the chief was a respectable looking man, with a large family by one wife. I found the young men here particularly curious and forward, and I only kept my temper and got rid of their annoyance, by making a party to fish in some large pools half way to our last halting place.

We stripped, waded, swam, and hauled the sein, and got plenty of fish.

Among other fish caught here were two which seemed to be novel: one, eighteen inches long, was brown on the back, with red blotches on the sides, and yellowish white belly; it had a purse or bag-like mouth, and eleven rays to the dorsal fin, was evidently a barbel (barbus), but peculiar from having its nose produced and rounded, like the Cyprinus Narus, and from the front of the back being elevated and rounded.

The other was a foot long; its back was bluish;

yellowish on the sides: it was probably a Leuciseus, for there were no indications of beards. Mr. J. E. Gray, of the British Museum, to whom my sketches were shown, proposed to call the first of these two varieties of fish, *Barbus Namaquaensis*, and of the second, he said that he was not aware that any species of the genus, to which it appeared to belong, had before been recorded as a native of the southern part of Africa.

At Kuis we were fortunate in adding to our collection of birds many fine varieties. Among others were a very fine eagle, with a red breast and white tail, pheasants, black and white spurwings, brown and black hoopoes, woodpeckers, kingfishers, white egrettes, large brown geese, water hens, &c. A collector would be well rewarded, if, associating himself with the Buys, with Aramap, or with any of the other good people of the land, he proceeded up the Fish river from its mouth to its source, then turned to the right to look at the Nosop, and to the left to obtain some large skins at the Bull's Mouth Pass.

New birds from all parts of the world are continually being described, but preserved objects of natural history are little thought of in these times compared with living specimens, there is such a mania for them at present, and for the formation of zoological gardens. I was unfortunate with my young gemsbok, as well as with the springbok, ostriches, &c. Henrick the driver carried the gemsbok for a month before him on an ox, at last it died from struggling to free its legs; it seemed always stupid and untameable, and was totally different from the pet-springbok in disposition.

Though there is no gemsbok now in England that I am aware of, not a living specimen of this antelope, which can contend successfully with the king of beasts, yet I was not sorry when my gemsbok died, for the valuable servant who carried it was then saved a great deal of annoyance.

CHAPTER IX.

Despatch a Messenger to the Chief Amral to prevent War in the land—The Tribes jealous of each other—Lions prefer human flesh—The Misfortunes which befel a party of Hunters under Amral—Leave Kuis—Travel along the Kaikum River—Boschman-fishing—Dreary Plains—The Hartebeest—The Chup River—Departure of the Escort—The Kuhap River—Visit of Bechuanas to the Fish River—Discovery of a City in the wilderness—The Huntop—Dreams—The Poisoned Pool—Intense cold—Travel along the Summits of the Bulb Mountains—" Flesh is weak "—Distress of the Oxen—The Water-place of Kuma Kams—The Fountain of Blood—Descend the 'Un'uma Range—Khumees—A Hairy Shepherd—Halt to repair Damages.

ARAMAP, and his dependant, Kraai, had both mentioned to me that they expected soon to be engaged in fighting with the people of Amral, the chief of the Keikouas * Namaquas, who was not many days to the south-east of Kuis, (and who had just been bitten by a lion,) and they requested that I would interfere to prevent war in the land. I said I should do my utmost to effect this; and, accordingly, I now despatched a mes-

* Great mens' partners.

senger to Amral with a government medal and a letter, which the messenger got by heart to communicate its contents to the chief.

The sight of a written paper has a great effect in Namaqua land.

I explained to Amral why I had come into the land—to see it, and to ascertain if a trade could be opened with the natives. I regretted that our route had not conducted us by his kraals, and that he had met with an accident from a lion, which 1 trusted he would soon get the better of. I assured him that no boors would be allowed to come into Great Namaqua land to settle in it, but that some traders would probably come to supply the Namaquas with what goods they wanted, and that then one or two missionaries would follow to teach the people, but for these great advantages it was necessary that *peace* should be preserved in the land.

I stated that I had heard of some differences and disputes between his young men and those of Aramap and Kraai, but 1 hoped that he (Amral), as a good and true man and a great chief, would use his influence to make his people live in good fellowship with the others—that I was on my way to the Cape by Komakas (Mr. Schmelen's) place, and that I now sent a token of the king my master, which was given to great captains in proof of friendship.

This message had, I believe, the effect which it was intended to produce. Amral himself was peaceably inclined, but his young people were restless, and anxious to try their strength with those of Aramap and Kraai, their neighbours. Perhaps, they were also jealous of Aramap's success, and envied the fine country he had acquired, though the Keikuas had also plenty of game in their country, and had range enough.

During our journey from Niais to Kuis, I asked why we saw so few people on our route? and the answer was, " There are plenty of Damaras on the hills, but there are no people on these plains at present, the last who were here were eaten up by the lions."

" Surely, not all of them?"

" No, not all, but a great many were eaten, and the rest ran away; for the lions, when they once taste human flesh, will leave all other game to hunt men, and will leap over a fire to get at them."

What had just befallen Amral and some of his people confirms this. The chief, and 200 of his men, had left Koonhop, his usual lay-place, which is two days west of the Nosop River, and were proceeding up the Nosop, hunting with guns and bows, and were also, perhaps, intending not to come home before securing some Damara cattle, when the following accidents happened :—

The first day, some of the hunters were pursuing an elephant in the river, when they came across lions, which pursued them, and they only saved themselves by abandoning a horse, which the lions devoured. Three of the party then made a scherm or screen of bushes opposite a pool, where they expected elephants or rhinoceroses to come and drink, and inside the scherm they dug a hole, the better to conceal themselves. Rhinoceroses came, two of the men fired and missed them, and the third was about to fire, when a lion, now first observed, sprung in upon him, and carried him off, without his companions being able to help him, and all that they afterwards saw of him was one leg.

Next day one of the Keikouas wounded a rhinoceros, which turned and charged him, came on with its horrid blast and scream, tossed him in the air, cutting his leg severely; his gun fell one way, and he another. He was picked up very lame, and sent back to Koonhop on horseback.

Immediately after this, two Namaquas and a Boschman of Amral's party, were sitting by a fire at night behind a scherm, when a lion came, seized one of them, dragged him through the fire, and bit off his back; one of the men fired, but missed, on which the lion dropped its dying victim, and growled across the fire at the two men, they durst not repeat the shot; the lion then took up its prey in its mouth, and went off with it.

Alarmed at these disasters, the Namaquas now assembled in one large scherm, and sent a Damara slave out at night for water. He had no sooner reached the pool than he was seized by a lion, he called in vain for help, and was slipped off among the reeds, and next day his skull only was found, clean licked by the rough file of the lion's tongue.

Amral and his men now turned out to hunt lions only, and in proceeding on some spoor up a hill, they soon saw two lions making off among the long grass and bushes; those of the party who had horses mounted them, and were soon able to shoot one of the lions in a bush; they then sat down, ate and refreshed themselves; and in the afternoon went after the largest and most savage of the two lions, he who had probably done all the mischief previously by devouring three men.

The lion had ensconced himself in a patch of reeds, which were set fire to on the windward side, and as he came out before the flames, one man fired at him and missed him, and he was going off in the most deliberate manner, when two better marksmen struck him in the loins and fore arm, which seemed to shake him, another man's ball then struck short, entering the ground before the lion, when he immediately turned and charged, with a loud roar, in the smoke.

"Death poured from his eyes along the quailing bands. His joy was in the fall of men. Blood to him was a summer stream, that brings joy to the withered vales from its own mossy rock."

The hunters sat down in a moment, and two of them fired but missed, when the monster dashed in amongst them, and seizing Amral's brother by the back, he tore out his ribs and exposed his lungs. Amral rushed to the assistance of his unfortunate brother—his gun burnt priming, and throwing it down, he seized, in his desperation, the lion by the tail to make him let go his expiring victim; the lion did so, and turning on Amral, with a stroke of his paw he grazes his forehead, tears a large piece of flesh off his left arm, and wounds his left hand, Amral springs back, when the lion strikes him again on the side and throws him down, Amral quickly rises, when the lion, intent on his revenge, and smarting under his own wounds, fixes his claws in the sash of Amral, and gets one of his knees in his mouth, Amral falls, the lion then sits over him, mumbling his left arm; Amral, torn and bloody, calls in a feeble voice to his people, who were round him, but at a little distance, to shoot the lion from behind, and one of them destroyed him with a ball through the brain. The dead body of Amral's brother was then taken up and buried, and the hunters, after thus losing four of their number at the commencement of their expedition, thought it was as well to give it up, and they returned home, bearing their wounded chief.

This is a specimen of what may be met with in the region of lions; there, are danger and excitement enough for the keenest Nimrod, and plenty of what the Americans would call " rough and tumble."

Time pressed, and as I could not go out of my way to visit Amral, on the 13th of June I left Kuis, and on doing so we immediately exchanged the good country to the east of the 'Oup river for a barren waste, to the west of it. We halted at Kaikum or Pack river, twelve miles, and our course was now S.S.W.

Next day we packed off after eleven miles, along the same river, at two or three shallow pools, where we saw two Boschmans catching small fry, in this simple fashion :—they collected a quantity of grass, and laid it across the water at one end of the pool, then wading in, they pushed the grass before them to the other end, when the little fish leapt out in great quantities on the wet sand.

Our march on the 24th was over plains of vast extent, desolate and silent as the grave, covered with stones and scattered bushes—the sight of these solitudes was enough to damp the ardour of the most determined voyageur, and to sink the spirits, as the eye wandered over the waste in search of some hill, or some other object on which to rest; the next feeling was to hurry across them as rapidly as I had formerly done on a low troitski, with three long-tailed galloways, over the Russian Steppes, and to try to get over their weary extent and to their termination; but speed was impossible—" stick to the trek (or line of pack oxen) and the trek will stick to you"—so it was necessary to move slowly, to bring on all our cattle and stores.

After twenty-one miles we packed off at

Nubapis or Rhinoceros water, also on the Kaikum river.

That fine animal, the Hartebeest, which I
had not seen since I had been on the banks
of the Keiskama, in Cafferland, now appeared,
to enliven us.   It is one of the best known of
the antelopes of South Africa, and the flavour
of its venison is much esteemed.   In size the
hartebeest is equal to the largest of the deer
tribe, its horns are thick and annulated, rise
nearly straight from the forehead, and then
bend suddenly backward.   When brought to
bay it is said to drop on its knees and then rush
forward on its enemies.

On the 26th travelling due south, we passed
the Atkuma river, whose course was easterly,
towards the Great Fish river, and still traversing
the same cheerless solitudes, where, if a person
had strayed, there was no visible means of sustenance for him.   We packed off at the Chup
or do nothing river.

Choubib, from some news he had heard at
Kraai's place, now became anxious to return
home, and Kuisip, who was always led by

Choubib, also asked if I could now spare him, for a direct road led from the Chup to the Kamop river, where their people were, and the road to Komakas, which I intended to follow, was out of the way for them. As I had Henrick Buys and three or four of his men who knew the country, to guide me, and assist in driving the cattle, and as we hoped to be able to defend ourselves against Boschmans and lions, I had no occasion for the further services of the Koutoukooas, or " short backs" of Kuisip. From the chief Kuisip, I was sorry to part, for he was a good man, and was always anxious to assist me with his people; but old Choubib had long disgusted the whole party with his bad temper and covetousness; whenever a case was opened he used to come prying into it to see what he could ask for out of it; my men were all delighted to part with him, but it was done in good humour.

I gave to the chief and to Choubib, according to agreement, a new musket each, the value of of which was eight oxen, I also gave them fourteen pounds of lead, and seven pounds of powder each, shawls, handkerchiefs, tinder boxes, &c.

also to each of their men a small present, and
they went on their way contented, with some
sheep from my flock for their support on the road.

No sooner were they gone than a fine Damara slave boy of Choubib's, Apollos by name,
appeared among my people, he had run away
from his master, and wished to go with us;
Choubib sent two men after him, who carried
him off by force, and before we could interfere,
they flogged the poor fellow most unmercifully
with thorn bushes. Apollos was the son of a
chief, and his bearing was always very independent.   I asked Saul if he would like to go back
to Choubib's people, and he began to cry, fearing that I would send him again to climb gum
trees.

We continued our journey crossing the Kurie
Ku, or Strong Running river, and packed off at
the Kuhap or Knot river, (twelve miles,) with
a lighter and more manageable party.   We had
thunder and rain at the Kuhap.

I said before that the chief Aramap had told
me that it was impossible to get to Latakoo
from the Fish river, without going nearly along

the course of the Orange, and that none of the Namaquas knew a shorter route across to the Bechuana country, but it appears that in former times the Bechuanas knew and used a short route to the Fish river, for Henrick Buys re- membered when he was a youth of about fifteen (thirty years ago,) to have seen on the Fish river, and east of where we now were, a party of Bechuanas (wearing handsome karosses, or mantles of blue cat, jackal, and other skins, and a cloth (?) round their waist, the ends of which were brought up between their legs) who came to exchange their small axes with the Namaquas for cattle, and these people had come directly across from their own country without going near the Orange river. But this was not the last time, he was aware that some of the eastern tribes had visited the borders of Namaqua land. For a few years ago he had gone with a party, due east from Bethany, across the Fish and Nosop Rivers, and had got to a branch of the latter, where they expected to find plenty of elephants, when they descried on the plain be- fore them, to their surprise and fear, a town of mud huts. They concealed themselves for a time to observe what passed in the town, but seeing no people in or about it, they took courage and entered it.

The huts were small, capable of containing only two or three people each, and were constructed of wattled stakes and clay. They were arranged close together in a large circle, outside of which was a fence of bushes, and inside were numerous cattle kraals also of bushes, disposed like figures of 8 connected longitudinally. Pieces of pottery and cloth were lying about, and it was evident that this " city in the wilderness " had been inhabited by a great multitude of people, who had not long abandoned it.

A Boschman, with whom Henrick's party fell in, said, that the town which they now saw had been raised by a large tribe, who had come from the eastward not long before, and who were called *Manchatees*—that some time after they had settled themselves in their town, the Kamaka Damaras come down on them from the north, and that a great battle had taken place, in which the Damaras were defeated, and then, after a short residence at their town, the Manchatees had gone away towards the N.E., and the Boschman saw them no more.

In Thompson's excellent work on South Africa, he describes the invasion of the Bechuana country, in 1823, by an immense horde which came from the eastward, driven from their own country by the warlike Zoolas, and called *Mantatees*. After being defeated by the Griquas, who, mounted on horseback, attacked them successfully with their guns, near Latakoo, the horde was scattered, and doubtless it was a division of it which had wandered as far as the tributaries of the Nosop River, and which had constructed the town which Henrick Buys had seen.

On the 18th we continued our journey, two zebras were shot, and after a march over a level country of sixteen miles, we were refreshed with the sight of the constant waters and green trees of the Huntop, or Off-running river, under a deep bank below us. We zig-zagged down the steep, and gladly packed off, on the anniversary of glorious Waterloo.

We halted a day here, to fish in the deep pools, and we caught barbel of a similar description to those noticed at Kuis. A dish of fish was a great treat to us, tired as we were of daily flesh, without bread, vegetables, or salt. I was still quite well myself, but two of my white men were much disordered from indulging in the yellow fat of the zebra; however, two or three strong doses of medicine brought them round again.

I felt now a constant longing for bread, which it was difficult to get the better of, and I dreamt at night of rioting in the midst of goodly loaves in bakers' shops, but, alas! in the morning we had nothing for it but *revenir à notre mouton.*

We made a short march down the Huntop of seven miles, and then leaving its pleasant banks, we travelled for seventeen miles over a dreary country, and with the prospect before us of a poisoned pool. We halted at a place called Khoons, or ground, on the Keisu, or black pot river, and I ran down to the river's bed to see if all the water was poisoned in the pools, at which I had heard that Aramap had just lost some horses. But

before I could stop them, half of the bullocks
ran past me and plunged into the water, and
drank. I thought they must surely die, when I
discovered that there had been two pools, an
upper and a lower, and that the lower pool, at
which the cattle were drinking, had been fenced
round with thorns by the Boschmans, and
that in the dried-up bed of the upper pool,
which had been left unenclosed, there lay several stems of the *euphorbia candelabra*, kept
down by stones, the poisonous juice exuding
from which, had rested like a scum on the surface of the water to poison zebras. Most unfortunately twenty horses, which Aramap had
procured from the south, had on their journey
to Niais, drank of the poisoned water, and had
all died; their remains lay not far from the
poisoned pool.

It is not particularly agreeable after a long
day's march to come to a pool, about the whole-
someness of whose water one is quite uncertain,
and where, if the water (as it usually happens to
be) is thick, it is impossible to see whether it
had been drugged or not. The only remedy in

a case of this sort is to feel about in the bottom of the pool with sticks, and with one's naked feet, for the prickly stems of the euphorbia, or to skim the water carefully, and to use as little of it as possible, when there is any doubt about it.

We now travelled west for twenty-one miles, saw numerous traces of the hartebeest, and on our right were some of the off-sets of the 'Un'uma or Bulb mountains; we began to ascend, and off-packed at a dry place, where we could not sleep all night for the cold, and where (it being now the middle of the South African winter, and our elevation very considerable) our karosses were stiff with hoar frost.

The bullocks were packed with difficulty, with benumbed fingers, and shivering in our skin clothes, we ascended on the morning of the 23d, by a precipitous road to the table land, on the tops of the 'Un'uma mountains; here we saw a troop of half-a-dozen cameleopards, and, after thirty miles of a cold walk, we off-packed at Kurumie, or blow mouth water-place, on the Kutip, or Hammer river.

Here we dug a hole ten feet deep—but we

could not keep it free from sand, to allow the cattle to drink, and Henrick the driver, was nearly smothered by the sand falling in upon him—neither could we get fire-wood to cook our victuals or keep ourselves warm. We were also beginning to leave behind us some bullocks overcome with fatigue, and had not a promising prospect before us in any way, till we should reach the Koanquip.

We had been highly favoured hitherto in having lost so few cattle, considering the great extent of country we had traversed; and it was incumbent on us to be very thankful for the manner in which we had been dealt with; still "flesh is weak," and it was impossible to keep one's spirits up at all times, particularly when we were again in great distress for water.— "This time next year," we said, "we may be in a better position."

On the 24th, a march of eleven miles brought us to the Ku''ums, or knot mouth river; here there was no water for the cattle, and only a dirty puddle under some stones for the men— strange to say, it was now the fourth day since

L 3

the wretched oxen had tasted water, as the Poisoned Pool was the last place where they had
drank.   I expected to see the whole herd lie
down and perish before my eyes every instant,
no European cattle could have held out in the
inconceivable way these hardy Africans had
done.

I heard that there was a water place some
distance from the Ku"ums, and to prevent any
carelessness about the cattle and sheep, I set off
with them on foot over the heights to the S.W., but
it was not till after a rough march of ten miles
that we came to a very remarkable pool, below
what was occasionally a cataract; high cliffs shut
in the pool, which was deep and black, but the
water was of excellent quality.   Some armed
Boschmans ascended the rocks as we drew near,
and the impatient cattle were driven to the water
in squads, to enable them to drink with the least
inconvenience after their dry journey from the
Keisu of seventy-two miles.

This water place was called Kuma Kams, or
the water of the beast tribe, and near it was a
heap of stones, eight yards long by one and

a half high, in a cleft between two eminences, which the Namaquas said was a heap over their deity, Heije Eibib.

I got back to the Ku''ums late at night, my people thought I was lost, and had made great fires, to direct me to the bivouack. After a day's halt to refresh the cattle, we continued our journey along the cold and flat summits of the Bulb Mountains, and turned aside to get water at the fountain 'Ahuas, or blood. In this was said to dwell a snake which guarded it, but strange to say, when the fountain was reached, it was found to be dried up, and a water snake, about six feet long, brown above and yellow below, lay dead beside it. The Namaquas immediately cried out, " Some one has killed the snake of the Fountain, and it is therefore dried up."

Not far from the Fountain of Blood, a young Boschman and his wife were met, and the woman accused her husband of having committed a great crime; she said, that the day before, they had drank at 'Ahuas, and the Boschman seeing the snake there had killed it. He excused himself by saying that he was a stranger in that part of the country, and did not know that the snake he had killed at the edge of the water, was the snake of the Fountain.

'Ahuas was not the only fountain in Namaqua land which was superstitiously believed to be preserved by a snake; but it was the only one which had come in our way. It was singular enough that it should have dried up immediately after the death of the snake; perhaps a hole which the snake made in the soft mud might have kept the fountain open.

We descended by a very rugged path, and in three quarters of an hour, from the summit of the 'Un'uma range to the plain below, and then reached the junction of the Great and Little Koanquip rivers, twenty-seven miles, where we lay among some bushes during a very cold night.

Passing along the Great Koanquip, some remarkable summits of the 'Un'uma arrested our attention; they stood boldly out from the range, and rose like pyramids or truncated cones into the sky. The first of these mountains I named

after my brother; another, opposite the Khumees or Lion water place, where we halted, after
W. D. Cooley, Esq. F.R.G.S., who proposed the
Delagoa expedition; and a third, after a valued
friend in the East, the Honourable Colonel
Morison, C.B., Member of the Supreme Council
of India.

Khumees, as its name implies, is a favourite
haunt for lions. We saw none there; but a new
pheasant was noticed among the bushes, with a
white ring round its neck, a brown back, and a
long tail; also plenty of baboons were seen on
the rocks above the water.

The Namaquas said, that, not long ago, a man
had brought up a young baboon, and had made
it his shepherd. It remained by the flock all
day in the field, and at night drove it home to
the kraal, riding on the back of one of the goats,
which brought up the rear. The baboon had
the milk of one goat allowed to it, and it sucked
that one only, and guarded the milk of the others
from the children. It also got a little meat from
its master. It held the office of shepherd for

twelve moons, and then was unfortunately killed in a tree by a leopard.

I cannot vouch for the truth of this story, but we have instances at home of what ourang-ou-tangs can be made to do, and possibly the tale of the hairy shepherd may be quite true.

We were visited by Christian Buys, who lived not far off; he was a quiet respectable-looking man, and was dressed for the occasion in a sol-dier's red jacket, which he had given sundry oxen for at the sea. Two more marches brought us to Henrick Buy's people, who were tarrying on the banks of the Koanquip. Here we halted to make and mend clothes, and were glad to get again under cover, for a time, for the ther-mometer, for several mornings, was at 32°, with plenty of hoar frost.

" Dissolve frigus, ligna super foco
Largé reponens, permitte divis cœtera."

CHAPTER X.

We had not been long among Henrick Buy's people, before his elder brother, "oud Jan," rejoined us: he had got safe to the banks of the Koanquip with the waggon, which I now recovered.

After Jan left us on the Kuisip, he got to Ababies, where he saw no Boschmans. It was at Ababies, if any where, I thought he would have found the water poisoned (in revenge for the theft of the Boschmans' skins), but it was not so. He then got the waggon through the pass

of the Bull's Mouth with less difficulty than
before, as the rhinoceros hide was the only thing
of any weight in it.   He halted at the place
where the first rhinoceros was killed, and the
oxen rushed into a deep water hole, out of which
they were hauled by dint of great exertion with
stout thongs, with the exception of one of my
best, which was drowned.   At the entrance of
the pass, the oxen betrayed alarm on approaching
our old scherms there, when Jan going forward
to see what was the matter, observed a large lion
gnawing a bone in one of our lay places.   The
lion slowly retreated up the hill, and lay down
to watch the party, Jan, knowing the danger he
would incur if he stopped as he had intended, at
the entrance of the pass for the night, merely
gave the oxen water, and hurried on to a
safe distance, and without further adventure
he crossed the Great Flat, and reached the
Koanquip.

I was very glad to see Jan again, and was
much obliged to him for the trouble he had
taken on my account, and he and Henrick now
said, that they would assist me with oxen as far

as Komakas, or even to Cape Town, if I liked, and I gladly availed myself of their offer, for many of my cattle were very thin, and much distressed.

The waggon, shaken by very rough travelling, was repaired, and the wheels were laid in wet sand, to tighten them, and where the wood would not meet the iron, wedges were introduced. The women were busily employed in making and repairing our skin clothes, we now got rid of all our rags, and the party turned out in very decent order, with leather hunting frocks, good trousers, and untanned shoes.

But all this took up a fortnight, during which I took my exercise under a remarkable cliff of the 'Un'uma, like a baronial castle on a hill.

We were not much annoyed with wild beasts at the Koanquip, only one night a hyena came to where we lay, and took up an iron kettle, in which there had been milk, and carried it off. In the morning it was nowhere to be found, till on our observing the foot-prints of a hyena at our fire, they were followed, and the kettle, the lid of which was fast, was found under a bush, half-a-mile down the river. A greasy iron pot-lid afterwards disappeared, but that was never recovered; it was probably buried.

The baboons sometimes annoy the Nama-quas by stealing their children;—thus a fire-side story narrated, that the children belonging to a heis were playing at some little distance from the huts with bows and arrows; in the evening they all returned home, save one, a boy of five or six years old, who lingered behind, and was soon surrounded by a troop of baboons, which carried him up a mountain.

The people turned out to recover the boy, and for days they hunted after him in vain; he was nowhere to be seen; the baboons also had decamped from the neighbourhood.

A year after this occurred, a mounted hunter came to the heis from a distance, and told the people that he had crossed at such a place the spoor of baboons, along with the foot-marks of a child. The people went to the place which the hunter had indicated, and they soon saw what they were in search of, viz. the boy, sitting on a pinnacle of rock, in company with a large baboon.

The moment the people approached, the baboon took up the boy, and scampered off with him; but, after a close pursuit, the boy was recovered. He seemed quite wild, and tried to run away to the baboons again; however, he was brought back to the heis, and when he recovered his speech, he said, that the baboons had been very kind to him; that they ate scorpions and spiders themselves, but brought him roots, gum, and wild raisins, seeing that he did not touch the two first delicacies, and that they always allowed him to drink first at the waters; thus, apparently, acknowledging his superiority to themselves.

This, as the Americans would remark, is interesting, if true.

Our preparations being completed, and the waggon again loaded, Henrick the driver gladly cracked his whip, and we set out in good spirits for the south. There were, however, three impediments which it was likely we should meet with on the journey—Henrick the robber, and his myrmidons, the " leather shoe wearers," in revenge for having been obliged to surrender Choubib's cattle and sheep, and to get some of our powder, hearing of our return, might waylay and attack us—we were about to pass through the most dangerous field for lions in Namaqua land—and, lastly, the Orange river might be in flood. However, we had some good men with us, and we were resolved, at all events, " to die hard."

We reached Bethany on the 13th of July, fifteen miles. Here all the oxen ran away, and we were delayed for forty-eight hours, till they were brought back from the Koanquip. In the garden at Bethany the " Dakka rookers," or smokers of the intoxicating and deleterious leaves of hemp, had an opportunity of filling their pouches with what they preferred far above tobacco even, from the pleasant visions which dakka (like opium) inspires. Yet dakka, like the extract from the poppy, makes those addicted to it habitually stupid, and soon wears them out.

On the 15th and 16th, after crossing the Koanquip, we were at Piet Buys heis, consisting of five or six huts, on the Quahanap or javelin river, and then crossing the Koanquip again, we outspanned under a group of nameless hills, which I called after my respected friend, George Banks, Esq., formerly mayor of Leeds. Two lions followed us to Hudap, or Ground Path, but did not attack us. We were now in the " Sharp Lion Country."

To Ukanip, or bitter river, was twenty-two miles: here we made large fires for the lions, and were up repeatedly in the night to keep the cattle and sheep near us. We made a short march to get water, and then reached the Hoons, or turn round river, where some time before a poor Boschman, who slept here in fancied security, in a cleft between two limestone rocks, had been dragged out and devoured; his kaross and staff only were found. It was considered so very dangerous a place, that we durst not trust only to the dogs as heretofore, but I, with my own men, kept an hour's watch each, during the night, to prevent the fires going out. Lions came round us, but spared us on the night of the 20th, though I lost an ox on the night of the 23d at Hoons fountain.

Passing through a country without inhabitants, and abandoned entirely to the dominion of lions,

with hills generally of no great elevation, and
plains covered with prickly shrubs, we outspanned on the Kubieb, or Stick Grass river,
where was much iron ore. The lion field was
now succeeded by that of the leopard. We killed
and ate an ostrich, which, after being wounded,
did battle with the hunter; and we had now
daily the change of ostrich eggs to feast on.

Twenty-two eggs were the greatest number
we found in a nest, and we never got a bad
one. Mixing the eggs with a little fat, or occasionally blood (as we were not Jews in any thing
but the beard), we made omelettes, which were
not to be despised anywhere.

It is commonly supposed that the ostrich is a
very stupid bird, that when hard pressed it conceals its head in a bush, and because it cannot
see the hunters it imagines they cannot see it,
that it is careless about its eggs, &c.; but it appeared to me that the ostrich has quite as much
intelligence, and, with the exception of leaving
its eggs for some hours, in the heat of the day,
for the purpose of feeding—has as much care for
its offspring as others of the feathered tribe.

What befell Elliot about this time proves all this.

One evening, he came to me with his face flushed, and out of breath. " What's the matter now ?"

" Sir, I've had such a chase after a sick ostrich, and the beast got away from me after all, sir— it got out of a bush, and ran off lame of a leg, and with its wings flapping, for it was mortal sick or badly wounded. I did not stop to fire till I got close to it; two of the dogs and myself chased it to make sure of it—it lay down sometimes, and the dogs could make no hand of it; then it got up again, but so bad was it, that I thought it would tumble over and break its long neck every minute; but I ran three miles after the thief of the world, and it bothered me entirely."

I told him it must have been playing the same trick which partridges practice at home when they have eggs or young, viz., going off as if crippled, to allure the foot of the stranger from their charge. But Elliot maintained that the ostrich was sick or wounded, and could not help its limping off; till Henrick the hunter came up, carrying half a dozen eggs, and reported he had shot the ostrich which we were talking about.

" I saw it start," said he, " and Elliot after it; I looked about and found its nest with fifteen eggs in it; as it was near sundown, I knew it would soon come back to the nest after decoying Elliot to a distance, so I made a screen of bushes near the nest. I sat down behind it for half an hour, and shot the ostrich on the eggs."

At Herees or wet ground, the scene changed, it was now the South African spring, and though over the plains and on the hills there was the silence of the desert, interrupted only occasionally by the call of the bekmakeri (telophonus collaris), a well known bird at the Cape, and which seemed to welcome us back again, the landscape was most refreshing to look on. The beautiful green of the bushes was contrasted with the varied and resplendent colours of the wild flowers; the bulbs had burst their cerements and were now shooting up or spreading on the surface of the ground their strange leaves. As we walked after the cattle, and descried, with considerable delight, the mountains south of the Orange river, the feet of the oxen constantly crushed plants which gave out a pungent and aromatic odour. Thus were our senses regaled.

But as " the most lovely roses have thorns," so snakes and scorpions were now rife amidst the flowering plants; and in one narrow valley where we slept uneasily, owing to the overpowering and humid smell of the plants, scorpions ran about us in every direction. But I had plenty of eau-de-luce for their troublesome stings, and a small cylindrical air pump for puff adder bites.

Leopards, with their beautiful skins, lay among the rocks, but spared our sheep. However, when they do attack a flock, they are more destructive than any other wild animal. One leopard will kill twenty sheep in a night, and suck a little of the blood of each only; and when brought to bay in the field, it springs from one hunter to the other, and claws their faces. The experienced cover themselves with their karosses and sit down, and the leopard will then probably spring over them and pursue fugitives.

On the 30th of July, thinking of past scenes of pleasure, and anticipating future ones, I went merrily ahead to examine the ford of the Kunarusip, or that of the Ebony Black Sheep, where we proposed to cross the Orange river. I urged my ox across the wastes of sand which mark the vicinity of the great stream, and I passed under the grotesque hills, tossed in romantic and endless confusion, which enclose its channel; at last I saw its broad waters rolling clear (fortunately not turbid and in flood) between its ebony-covered banks, and I welcomed it as an old and valued friend.

But black clouds to the eastward, above the sharp peaks and ridges of the distant summits of the " Gariepean walls," warned us that there was no time to lose in getting across the river. Accordingly, slipping off our clothes, and with staves in our hands, we waded, with difficulty, over the smooth stones, and with the water up to our waists, across the upper part of a rapid, down which we were nearly hurried. The waggon

oxen supporting one another, dragged the conveyance across, after much trouble, but the loose
oxen were all carried down by the force of the
current, and some of the packed ones also swam
across. Hours elapsed before the sheep could be
got over on the following day, and it was only
done by " wooden horses," which were made by
cutting trunks of trees, eight feet long, and with
a peg on them for one hand; they were bestrode,
and were propelled across by foot and hand; a
sheep was then pushed into the water, was driven
before the " horse marine," and made to swim
the rapid.

We were obliged to remain on a sort of island
in the river, for two nights, but over which, as we
saw by the dry twigs and grass on the trees, the
water occasionally swept. Our situation, with
the lowering clouds east of us hanging over the
mountains up the river, was not free from anxiety,
for thus it might have happened with the Gariep.

> ————— " The glorious stream
> That late between its banks was seen to glide;
> Had now sent forth its waters, and o'er plain
> And valley, like a giant from his bed
> Rising, with out-stretched arms superbly spread."

"I remember," said oud Jan, "some terrible floods in the Koanquip which surprised us. The bed of the river, towards the mouth, was quite dry, when down came the water suddenly, and with a great noise, and covered most of the trees. A man and a snake would then be seen in a tall tree, and they did not molest each other, from fear of the torrent; or a man and a panther would be seen together. Many of the people who were living on the banks were swept away and drowned; and we, who were at a little distance, heard the roar, and saw the dreadful consequences of the flood with trembling."

"On the banks of the Gariep, too, I recollect," continued he, "that a honey seeker once climbed a cliff and loaded himself with honey; whilst he was securing his burden, he heard below him a roar, and looking down, he saw that the river had come down so as to separate him from his people on the other side; he accordingly cut a block of wood, launched it, and bestriding it, he tried, for a long time, to get into the stream, but he could not, for the eddy always drove him back; at last he struggled out into the middle of the river, but there the force of the current was so great that he was forced to yield to it; he was carried down, and was quite unable to help himself. He saw dead sea cows rolling down the river, and trees which had been torn up by the roots; he passed over the tops of others, and he went a week's journey on foot, in two days and two nights, on the log, subsisting on the honeycomb, which he had before him. At last he came to where the trees ended below 'Aris, and he thought that he should surely now be carried out to sea, when a lucky branch caught him, and he stuck fast; but he was so cold and benumbed that it was a long time before he could walk and get to his friends again."

We gladly left the ford of the Kunarusip,* as the river is a deceitful beauty, and moving along the stream for five miles towards another ford, called Numedamas, or the Shining Eye, we turned from the Great River, and travelled south; but we lost our oxen, for they moved off

* A conspicuous mountain opposite the ford I named after James Mackillop, Esq., late M.P. for Tregony, a valued friend.

to a great distance to pick up their food, having subsisted on nothing but the tops of bushes and ebony leaves for many days. Where we passed the Orange there was no grass, though there is always plenty in other parts of the course of the river.

Travelling between hills and over fresh and rank vegetation as before, we reached Ubib or Brack Place, where we out-spanned among tamarisk-trees. The jackals at night yelling round us brought on stories at the fire of these animals, of their cunning as compared with the stupidity of other beasts; and I trust it will not be thought trifling, if I give a specimen or two of these Namaqua nights' entertainments.

Once on a time, a jackal, which lived on the borders of the Colony, saw a waggon returning from the sea-side, laden with fish; he tried to get into the waggon from behind, but he could not; he then ran on before, and lay in the road as if dead. The waggon came up to him, and the leader cried to the driver, " Here is a fine kaross for your wife !"

" Throw it into the waggon," said the driver, and the jackal was thrown in.

The waggon travelled on, through a moonlight night, and all the while the jackal was throwing out the fish into the road; he then jumped out himself, and secured a great prize. But a stupid old wolf (hyena) coming by, ate more than his share, for which the jackal owed him a grudge, and he said to him, " You can get plenty of fish too, if you lie in the way of a waggon as I did, and keep quite still whatever happens."

" So!" mumbled the wolf.

Accordingly, when the next waggon came from the sea, the wolf stretched himself out in the road. " What ugly thing is this?" cried the leader, and kicked the wolf. He then took a stick and thrashed it within an inch of its life. The wolf, according to the directions of the jackal, lay quiet as long as he could; he then got up and hobbled off, to tell his misfortune to the jackal, who pretended to comfort him.

" What a pity," said the wolf, " I have not got such a handsome skin as you have."

Again;—A tiger (leopard) was returning home from hunting on one occasion, when he lighted on the kraal of a ram. Now, the tiger had never seen a ram before, and accordingly, approaching submissively, he said, " Good day, friend ! what may your name be ?"

The other, in his gruff voice, and striking his breast with his fore-foot, said, " I am a ram. Who are you ?"

" A tiger," answered the other, more dead than alive ; and then, taking leave of the ram, he ran home as fast as he could.

A jackal lived at the same place as the tiger did, and the latter going to him, said, " Friend jackal, I am quite out of breath, and am half dead with fright, for I have just seen a terrible-looking fellow, with a large and a thick head, and on my asking him what his name was, he answered roughly, ' I am a ram.' "

" What a foolish tiger you are ?" cried the jackal, " to let such a nice piece of flesh stand ! Why did you do so ? but we shall go to-morrow, and eat it together ?"

Next day the two set off for the kraal of the ram, and as they appeared over a hill, the ram, who had turned out to look about him, and was calculating where he should that day crop a tender salad, saw them, and he immediately went to his wife, and said, " I fear this is our last day, for the jackal and tiger are both coming against us. What shall we do?"

" Don't be afraid," said the wife, " but take up the child in your arms; go out with it, and pinch it, to make it cry as if it were hungry." The ram did so, as the confederates came on.

No sooner did the tiger cast his eyes on the ram, than fear again took possession of him, and he wished to turn back. The jackal had provided against this, and made the tiger fast to himself with a leathern thong, and said, " Come on!" when the ram cried in a loud voice, and pinching his child at the same time, " You have done well, friend jackal, to have brought us the tiger to eat, for you hear how my child is crying for food!"

On these dreadful words, the tiger, notwithstanding the entreaties of the jackal to let him go, to let him loose, set off in the greatest alarm, dragged the jackal after him over hill and valley, through bushes and over rocks, and never stopped to look behind him till he brought back himself and the half-dead jackal to his place again.   And so the ram escaped.

Lastly, a jackal and a wolf went and hired themselves to a man to be his servants.   In the middle of the night the jackal rose and smeared the wolf's tail with some fat, and then ate all the rest of it in the house.   In the morning the man missed his fat, and he immediately accused the jackal of having eaten it.   "Look at the wolf's tail," said the rogue, "and you shall see who is the thief."   The man did so, and then thrashed the wolf till it was nearly dead.

These stories are absurd enough; but they serve to show how "children of a larger growth" can be amused in the region of the Orange river.

On the 3d of August the waggon went on to Aneip, or wet foot; and I went out of the way with Jan Buys and two or three men, to see a hole which was supposed to be inhabited by Heije Eibib, or the devil, and was the wonder of the country.

We crossed the Chunap, or troublesome river, and on the side of a limestone hill, we saw a black pit, the mouth of which was twelve feet across. I immediately set fire to a few bushes and threw them in, and lowered a stone with a fishing-line, whilst oud Jan stood by in alarm, thinking that something terrible would fly out of this reported bottomless and mysterious pit, when behold! it was found to contain nothing, and was only fifty feet deep, and so the cave lost its reputation.

Whilst we lay at Aneip, I saw through my glass two men approaching us from the Orange river, and driving a horse before them: in one of them I recognized a Dutchman who had run away from the colony, it was said, for a murder, and who had married a Namaqua wife, and who lived in the Orange river. In the other, I recognised Martinus, the miscreant Bastaard, who might have got all our throats cut by spreading the report, as I formerly stated, of our having gone to Namaqua land to take the country, and enslave the people!

Not wishing to have any altercation with Martinus, I sent a messenger to him to say, that I knew what he had done; that notwithstanding his cowardly and wicked attempt to injure us, he might now pass on, but that I could not see him. Martinus vehemently protested his innocence, and said that if I did not see him he would be disgraced; but I was obstinate and would not, and again sent to tell him to be off at once, and he accordingly disappeared.

An ox was falling behind each day from fatigue, and when we got to the rocks of Nus, it seemed that we could not drag the waggon into the colony at all; and therefore, I halted two days, dispatched two men to Komakas on my freshest oxen, to ask help from my worthy friend, Mr. Schmelen, and slowly followed with the rest. Of the oxen we had eaten a span of 14, and lost another span by fatigue!

A night march of thirty miles, chiefly on foot, brought us to the grass flat of Kuras, abounding in puff adders and other dangerous reptiles. At Kama we came on our old spoor. At Ukribip, Paul Lynx, the one-eyed chief of the Orange river mouth, came with three or four mounted attendants to congratulate me on my safe return. At the Kowsie I was met by a party of Mr. Schmelen's people, with a fresh span of oxen. We joyfully crossed the boundary, partook of some bread from Komakas, found that it had no taste, as we were spoilt by having subsisted so long on strong animal food; and on the 10th of August we again found ourselves on chairs, and in a bed, at the mission station at Komakas.

I recovered from the Kamiesberg my large expedition-waggon, and being again able, with my government order, to get assistance from the Boors, I parted from Jan and Henrick Buys, who had done me such good service, which I endeavoured to repay with a couple of capital elephant guns, &c., and, after a week's halt, I again "took the road," having received, as before, the kindest treatment from the family of Mr. Schmelen. Ὁ Ἄνθρωπος εὐεργετος πεφυκῶς. "Man is born to be a doer of good."

After a couple of marches, I visited a natural curiosity—not another cave of Heije Eibib, but a fat Dutch woman, whose bulk would have considerably assisted to fill the hole we had lately seen. Catchee Vanderkniver was the name of this lusty dame. She had not walked for some ten or twelve years, her fat had so increased upon her. I found her sitting up in bed in the middle of the day, dirty with snuff and grease, and more resembling a hippopotamus under a quilt than a human being. Before her, and by the side of the door, was her arm chair, in which I was lost. On one of the arms was fixed an iron spoon, on which Catchee was in the habit of hammering marrow bones, and then scooping up the rich morsels with her delicate forefinger. In a choking voice she told me she was eighty-four years of age, and remembered a French traveller in these parts who had taken liberties with a female relative of hers, on which she had taken a stick to him, and belaboured him out of doors. I could well believe this, for her slaves approached the testy old lady in fear and trembling.

Following nearly the same route to the Cape as I had taken on the outgoing journey as far as Field Cornet Hanekom's, I there turned to the left, to pass through the well-watered and beautiful district of the Four-and-Twenty rivers, spent two days at the delightful villages of the

Paarl and Stellenbosch; and as I had taken up my pilgrim's staff in the finest season of the year, when the face of nature was fresh and green with verdure, and brilliant with wild flowers, so, bearded like a pard, and brown as a Berber, I reached Cape Town during the same vernal time, on the 21st of September.

I was in health and strength; and I felt very grateful for the manner in which myself and people had been mercifully preserved on an expedition, certainly the roughest I had ever undertaken. In justice to my people, I must say that they were animated by a good spirit during the journey; they were respectful and obedient, and showed no want of courage; and, above all, they generally submitted with patience to the privations and hardships, to which such an undertaking as ours was always exposed.

Besides bringing back my seven attendants in safety to the Cape, my papers, drawings, and collections were also fortunately entire. Great part of our journey, of nearly 4,000 miles, was performed on foot, the rest on horseback and on a bullock. For about a year we had not slept out of our clothes or shoes, and subsisted on the coarsest food and the dirtiest water. The objects of natural history (particularized in the Appendix) comprised skins of many of the large game animals, as the rhinoceros, lion, zebra, koodoo, &c. The bird skins amounted to three hundred and twenty specimens; and a *hortus siccus* consisted of about the same number of plants.

During the whole journey the rocks were found to be chiefly of primitive formation; granite, old red sandstone, mica slate (particularly about the Kuisip), quartz, with imbedded crystals of felspar, hornblende, &c. I brought away many specimens of copper, and some iron.

It is earnestly to be hoped that extended intercourse with the natives beyond the Orange river, will result from this expedition, for their benefit and for that of our colonists; and that from the Cape, the blessings of civilization and religion will proceed by degrees, towards the Mountains of the Moon. Neither discovery nor moral improvement can proceed rapidly in Africa; but we must be continually endeavouring to promote both of these important objects.

The attentions of my friends at Cape Town, soon effaced the recollection of most of the troubles and trials I had experienced, during my twelvemonths' life in the bush—the associate of wild men and wild beasts. I became gradually civilized again. To reclaim me entirely from the savage state, I fortunately married, and in three days after the ceremony, I was on board the Hindoostan Indiaman; and, after a prosperous voyage of eight weeks, we once more reached London alive and well, on Christmas-Day.

APPENDIX.

No. I.

NOTICE OF THE OBJECTS OF NATURAL HISTORY COLLECTED DURING THE EXPEDITION.

AFTER my arrival in England I submitted the collection I had made of African quadrupeds, birds, plants, &c., to various scientific individuals, that I might have the benefit of their advice in giving a description of the most interesting specimens, and I was fortunate in having the assistance of gentlemen of such eminence as Sir John Herschel, Professor Lindley, Professor Don, Dr. R. D. Thompson, Mr. W. Herbert, Mr. W. Ogilby, Mr. J. Gould, Mr. J. E. Gray, Mr. G. Waterhouse, &c., whilst arranging my materials for this notice.

I. ZOOLOGICAL SPECIMENS.

Mr. W. Ogilby furnished the annexed list of new and rare mammalia collected during the Damara expedition.

I.—QUADRUMANA.

1. *Cynocephalus Porcarius* (Boddeart).

II.—CHEIROPTERA

2. *Nycteris Affinis* (Dr. Smith).

III.—INSECTIVORA.

3. *Chrysochloris Damarensis* (Ogilby) new species.
4. *Macroscelides Alexandri* (Ogilby) new species.
5. *Macroscelides Melanotis* (Ogilby) new species.

IV.—CARNIVORA.

6. *Gulo Capensis* (Schreber).
7. *Mustela Zorilla* (Desmarest).
8. *Viverra Felina?* (Thunberg).
9. *Herpestes Melanurus* (Dr. Smith).
10. *Cynctis Ogilbii* (Dr. Smith).
11. *Proteles Cristata* (Penny Cyclopædia, 1, 2).
12. *Canis Megalotis* (Cuvier).
13. *Canis Mesomelas* (Erxleben).
14. *Felis Leo* (Linnæus).
15. *Felis Nigripes* (Burchell).

V.—RODENTIA.

16. *Bathyergus Damarensis* (Ogilby) new species.
17. *Graphyurus elegans* (Ogilby) new species.
18. *Geosciurus Capensis* (Dr. Smith).
19. *Lepus Rupestris* (Dr. Smith).

VI.—PACHYDERMATA.

20. *Equus Zebra* (Linnæus).
21. *Rhinoceros Africanus* (Desmarest).
22. *Rhinoceros Simus?* (Burchell) an imperfect skull.
23. *Hyrax Capensis* (Schreber).

VII.—RUMINANTIA.

24. *Antilope Euchore* (Forster).
25. *Antilope Tragulus* (Forster).
26. *Antilope Traguloïdes* (Ogilby) new species.

It will be observed from the preceding catalogue that nearly a fourth of the mammalia are new species, that is to say, six out of twenty-six, and of the remaining twenty, seven, viz., Nos. 2, 9, 10, 11, 12, 20, and 21, are still rare in cabinets of natural history.

Besides the above, I brought home the first specimen which has appeared in Europe, of the *Aigoceros Niger* (Harris), or Sable Antelope. This magnificent animal has now been described in the Transactions of the Zoological Society, and was shot in the Zoolah Country in 1837 by Captain Harris, of the Bombay Engineers, an intelligent and enterprising officer.

" On examining the interesting collection of birds brought home by Captain Alexander from the interior of South Africa," says Mr. Gould, " I find many rare species, and several which appear to me to be new to science. The collection consists of 320 specimens and 125 species,—the following is a slight enumeration of them :—

" Of the raptorial order, or birds of prey, there are thirty-four specimens and sixteen species, viz., two vultures, *Neophron Peronopterus*, and *N. Monachus;* twenty-two falcons (ten species), among the more remarkable of which is a very beautiful eagle, with a red breast and white tail, and a very diminutive true falcon, half as large again as a sparrow, with a red back and spotted tail; eleven owls (four species), one of the genus *Surnia*, two of *Scops*, and one of *Athene*.

" Of the Insessorial order, or perching birds, there are two hundred and thirty-five specimens and seventy species —of these, seventeen specimens (six species) belong to the Fissirostral tribe, viz., one *Caprimulgus;* two species of *Coracias*, one of which the natives say alights on the horn of the rhinoceros. These examples are highly interesting, as showing the southern limit of the range of this beautiful tropical form. Two species of *Merops*, and one *Alcedo*. Of the Dentirostral tribe there are sixty specimens (twenty-two species), among which occur examples of the following genera: *Lanius* (two species), *Crateropus, Petrocincla, Saxicola, Ixos*. Of the Conirostres eighty-six specimens (twenty-five species), comprising examples of the genera *Euplectes, Estrilda, Amadina, Ploceus, Pastor, Lamprotornis*, and *Corvus*. Of the Scansores there are forty-two specimens (twelve species); of the genera *Colius, Bucco, Picus*, a very rare *Corythaix*, described by Dr. Smith, and an apparently new *Agapornis*, or small parrot. Of the Tenuirostres thirty specimens (six species), of the genera *Upupa, Rhinopomastus, Cinnyris*, &c. Of the Rasorial order there are eight or ten species, among which are two or three species of pigeons, and examples of the genera *Otis* (three species). *Cursorius, Pterocles, Francolinus, Struthio, Charadrius*, &c.

" The collection is not so rich in birds of the Grallatorial and Natatorial orders, as in those of the preceding. Among the grallatores are examples of the genera *Numenius, Ardea, Tringa, Gallinula*, and *Umbretta;* and in the Natatores, *Anser, Anas*, and *Podiceps*."

Mr. G. Waterhouse thus describes the birds which are figured.

1. Crateropus bicolor.
2. Petrocincla brevipes.
3. Lanioturdus torquatus.
4. Coracias nuchalis.
5. Psittacula roseicollis.
6. Francolinus adspersus.

Of these, 2, 3, and 6, are new species.

FAMILY MERULIDÆ.

1. CRATEROPUS BICOLOR, *Jardine*.

" Dirty white; wings and tail dusky black : legs and beak black.

	in.	lin.
Total length . . .	9	9
Length of tail . . .	4	7
——— of beak . .	1	2
——— of tarsus . .	1	4

Habitat, Mountains of Victoria, Damara land.

FAMILY MERULIDÆ.

2. PETROCINCLA BREVIPES, *Waterh.*

" Neck and back ashy-gray; crown of head gray-white, this colour extending on to the neck, where it blends with the deeper gray of the back; ear coverts blackish gray; primaries and secondaries black, edged with gray; under wing coverts, tail, rump, upper tail coverts, and the whole of the body beneath, from the lower part of the neck downwards, of a bright brownish orange colour: the two central tail feathers black; the remaining tail feathers slightly edged with black externally near the apex: legs and beak black.

	in.	lin.
Total length . . .	7	0
Length of tail . . .	2	10
——— of tarsus . .	1	$0\frac{1}{4}$
——— of beak . .	1	$0\frac{1}{2}$
——— of wing . .	4	3

Habitat, 'Tans Mountain, Damara land.

" The specific name of *brevipes* has been applied to this species, from the circumstance of the tarsi being so much shorter in proportion than in those with which it is most closely allied, I allude to the two South African species P. *perspicax* and P. *explorator*. It agrees in size with the former of these, but may at once be distinguished, not only by its short tarsi (which measure upwards of a quarter of an inch less), but in possessing a much longer beak, smaller and shorter toes, and weaker claws—the dilated portion of the central claw is not so wide, but extends nearer to the apex of the claw. The tail is longer in proportion, and the parts described as ashy-gray, are of a deeper hue than in P. *perspicax*, and are totally devoid of the blue tint which is conspicuous in that species. The whitish top to the head, and the dark ear coverts, also remove the present species from that with which it is compared. The P. *Explorator* is a larger bird, and differs so much in colouring and proportions, that it is unnecessary to make a comparison.

FAMILY LANIADÆ.

3. Genus LANIOTURDUS, *Waterh.*

" Bill as long as the head, moderately stout, slightly notched, and indistinctly curved at the tip; nostrils basal and linear, partially covered by small incumbent feathers; gape provided with slender bristles: wings short, feeble, and rounded; the secondaries large in proportion; first primary short; third, fourth, fifth and sixth nearly equal; the second shorter than the third: tail short and nearly even: tarsi long and somewhat powerful; feet short but strong; claws much compressed, and arched.

Sp. L. TORQUATUS, *Waterh.*

" General colour gray: head black; a frontal stripe, extending backwards over the eyes, white: throat, and fore part of the neck, white, the white extending in a narrow line round the back of the neck: a black belt encircling the back of the neck and extending across the chest: wings black; a large white patch at the base of the secondaries: the primaries white at the base, and edged with white at the apex; secondaries broadly tipped with white: tail white, the two central feathers having a large oblong black dash near the tip; vent feathers white, the white extending forwards along the centre of the abdomen: feathers of thighs broadly tipped with white: bill and legs black.

	in.	lin.
Total length . . .	6	0
Length of tail . . .	1	9
——— of tarsus . .	1	$2\frac{1}{2}$
——— of beak . .	0	11
——— of wing . .	3	6

Habitat, Bull's Mouth Pass, Boschman land.

" Not finding this bird described in the various works which I have consulted, nor yet being acquainted with a genus into which it can conveniently be placed, I have been under the necessity of characterizing a new genus for its reception. Its general characters indicate that its true affinities are with the Laniadæ, and it appears most nearly related to the members of the sub-family *Thamnophilina* of Swainson. Its thrush-like beak, short square tail, and long tarsi, however, will serve to distinguish it generally. It has the long and soft rump-feathers which are always found in the Thamnophili.

FAMILY CORVIDÆ (*Leach*)?

4. CORACIAS NUCHALIS, *Dr. Smith.*

" Brownish green: frontal feathers white, the white feathers extended backwards on each side, and forming a broadish line immediately above the eye: some bluish-white feathers on occiput: shoulders, primaries, secondaries, and tail feathers, deep ultra-marine blue; the outer web of the first primary greenish; the outer web of the other primaries are also greenish towards the apical portion of the feather: the two central tail feathers are dusky green, excepting towards the base: wing coverts purple-red, shaded into purple towards the base of the wing: tail coverts pale purple-blue: throat, chest, and body beneath rufous purple, each feather having a white line along the centre: beak black; legs brownish.

	in.	lin.
Total length . . .	13	0
Length of tail . . .	5	3
—— of beak . .	1	11
—— of tarsus . .	1	0

Habitat, Great Flat, Boschman land.

FAMILY PSITTACIDÆ.

5. PSITTACULA ROSEICOLLIS, *Vieillot.*

" Bright green: beak large and of a whitish colour; fore part of the head as far back as the eyes, of a bright crimson colour; a few feathers of the same hue narrowly surround the upper part of the eye, and occupying a small space behind the eye: the cheeks, throat, and chest, of a pale red, approaching to a rose colour; rump and upper tail coverts turquois blue: primaries with the inner web dusky:

tail with a broad band of a dull vermilion colour, beyond this, the feathers are green, becoming bluish towards the apex, and there is a black·spot on the inner edge of each feather: two central tail feathers green.

	in.	lin.
Length	5	9

Habitat, 'Un'uma Mountains, Great Namaqua land.

FAMILY TETRAONIDÆ.

6. FRANCOLINUS ADSPERSUS, *Waterh.*

" Above gray-brown; each feather crowded with extremely delicate and irregular black markings: primaries deep brown freckled with pale brown: neck and body beneath with irregular transverse black (or brownish-black) and white markings; each feather having from ten to twelve waved black bands, which are about equal in width to the interspaces: legs and beak red: tarsus short.

" In a second specimen, not adult, most of the feathers of the back present a white line along the centre, and an irregular, but somewhat lance-shaped black spot towards the apex; and on the belly most of the feathers are brown with transverse white markings, these white bands being finely edged with black: upper mandible black.

	in.	lin.
Total length . . .	13	0
Length of beak . .	1	1
———— of tarsus . .	2	0

Habitat, Fish River, Great Namaqua land.

" This Francolin, of which I have not seen the male sex, is closely allied to the F. *clamosus* of authors: it is however of a smaller size, and besides the difference in the colouring and markings, it may be distinguished by the tarsi being shorter in proportion."

———

Mr. Gray names a small lizard, with blunt toes, and which is said to be so poisonous that its bite occasions death within an hour, the " *Phyllodactylus lineatus.* Pale brown, temples, back, sides, and tail, with narrow longitudinal black streaks. Scales of the back, small, oblong, triangular, smooth; of the belly, rather larger and flatter; of the tail, larger, thicker, and more expanded. Inhabits Namaqua land, where it is called Geitjie."

Some land shells collected during the expedition about the Great Fish River, prove to be new to science. Mr. Gray thus describes them, the specimens now being in the collection of the British Museum, as is also the Geitjie.

Dorcasia, Gray, (*Helicidæ*).—Shell rather depressed, last whorle rather produced out of its direction, mouth orbicular ovate, elevated from the body whorle, quite free, oblique to the axis, and edged with a continued thin reflexed peristome; axis slightly perforated. (Animal like Helix?)

Dorcasia Alexandri, Gray.—Shell rather depressed; smooth, pale brown, pellucid, axis very slightly perforated. Diameter 15; axis 6 lines.

Helicodonta Sculpturata, Gray.—Shell, depressed umbilicated white; whorles with an irregular subcentral keel, and several spiral ridges, crossed with close narrow concentric ridges; mouth rhombic ovate, with a reflexed lip; throat with a strong ridge in the centre of the inner, and two equidistant ones on the outer lip. Diameter $3\frac{1}{2}$ lines.

Bulimus Hottentota, Gray. — Shell ovate ventricose, white, suture impressed; whorles rather convex, last much the largest; mouth ovate, not quite as long as the spire; lips thickened, not reflexed.

Bulimus Eulimoide, Gray.—Shell turitted; whorle pellucid, polished very finely, concentrically striated; apex rather blunt; whorles 8, rather convex; mouth, ovate; lips simple, axis scarcely perforated.   Axis 6 lines.

II.—BOTANICAL SPECIMENS.

"Among the plants," says Dr. Lindley, "collected beyond the country usually visited by strangers, occur two or three species of *Pappophorum,* a curious spiny plant, with hoary leaves and large flowers, somewhat resembling an *Escobedia* in size and appearance, but belonging to the natural order of *Solanaceæ*; several Acanthaceous plants, particularly one with bright blue flowers, and spiny leaves, allied to *Barleria* and *Acanthodium ;* the rare *Oloptera Burchellii*, a fine plant related to *Sesamum*, two trees belonging to different species of the genus *Ficus*, several *Amaranthaceæ*, and an apparently new species of *Aptosimum*.   The most curious plant is, however, what is called the *'Naras,* bearing a spiny fruit, double the size of an orange; but as yet it is impossible to form any decided opinion concerning it.   The specimen resembles *Schepperia juncea* so much, that it would be mistaken for it if it were not for the seeds, which are extremely like those of some cucurbitaceous plants."

On leaving the Cape in September, 1836, during the South African spring, and when the landscape is enriched with wild flowers of every hue, I collected first (according to Professor Don)—*Senecio sp., Lyperia sp., Vicia sativa,*

Dimorphotheca sp., Moræa tripetala, Arctotis sp., Passerina sp., Babiana stricta, Aristea cyanea, Ixia conica, Sparaxis tricolor, Ornithogalum sp., Lobelia sp., Heliophila sp., Aristea melaleuca.

Then about the Piquet berg I found—*Moræa Pavonia, Gladiolus alatus, Heliophila sp., Othonna sp., Mesembryanthemum pomeridianum, Crotolaria sp., Gladiolus sp., Babiana sp., Erica sp., Sparaxis sp., Trichonema sp., Solanum sp.* About the Cedar Mountains varieties of *Mesembryanthemum, Atriplex, Senecio, Cullumia, Anchusa capensis.*

At the Kamiesberg in October—*Gladiolus* (new), *Ornithogalum sp., Hebenstretia sp., Tritonia sp., Hemimeris sp., Homeria miniata, Pelargonium sp., Othonna leptophylla, Malva, Pharnaceum sp., Sutherlandia sp., Diosma sp., Penæa sp., Dianella sp., Anthericum sp.? Eriocephalus sp., Amellus sp.? Helichrysum canescens ;* and at the Orange River mouth—*Salsola sp., Mesembryanthemum sp., Tanacetum sp., Salsola sp.,* Ebony, *Rhus sp.*

At the Kowsie river—*Tamarix sp. ;* at Silver Fountain —*Pteronia sp., Othonna trifurcata, Salsola sp.* (soap bush), *Rhus sp.* (bush tea); at Hencrees—*Leyssera gnaphaloides ;* at the Warm Bath, in December—*Taxanthema sp., Acacia Giraffæ, Selago sp. ;* at Africaner's kraal—*Barleria sp., Pelargonium sp., Hibiscus sp. ;* at Karas Mountains—*Zizyphus sp., Acacia sp., Celastrus sp.*

About the Kamop river in February—*Tribulus terrestris, Aitonia capensis, Malva sp., Selago sp., Mahernia sp. ;* at Tuais, in March—*Indigofera sp., Cleome sp., Zygophyllum sp., Zizyphus sp., Justicia sp., Lantana sp., Sesbania sp., Clitoria sp. ;* on the Kei Kaap—*genus novum, Acacia sp., Cleome sp., Justicia sp., Malva sp.,*

Ocymum sp.; Great Fish river—*Relhania sp., Sesbania sp., Mahernia sp., Sesamum sp.? Bidens sp.*

Bull's Mouth Pass—*Acacia sp., Celosia sp., Cleome sp., Barleria pungens, Sesamum sp.? Cynanchum sp.* (the short and thick tree with the smooth bark, called "cheis" by the Namaquas), *Salvadora sp., Acanthus carduifolius,* fig trees.

At the Kuisip, in April—*Salvadora sp.,* willows, *Eragrostis sp., Arundo sp., Oplismenus sp.;* 'Naras, in May—*Acacia* (new), *Celastrus sp., Ruellia sp., Tragia capensis, Sesamum sp.?* great varieties of grapes, *Indigofera sp., Celosia sp., Periploca sp.* (?), *Cynanchum sp.*

At Niais, Damara land, in May—*Relhania sp., Clematis triloba, Tarchonanthus camphoratus;* at Glenelg's Bath, in June—*Justicia sp., Relhania sp., Acanthus carduifolius,* which, as we often saw its bright blue flowers on the dry plain, where there was nothing else agreeable to look upon, I termed " our comfort in the wilderness."

At the Huerap river—*Sida sp., Ficus sp., Barleria pungens, Euphorbia sp.;* Kubieb river, in July—*Rhus sp., Senecio sp., Heliophila sp., Dimorphotheca sp.;* Valley of Scorpions—*Sutherlandia sp., Selago sp., Galium sp., Dimorphotheca pluvialis, Didelta spinosum, Mesembryanthemum sp., Lyperia sp., Didelta carnosum, Leyssera gnaphaloides, Mahernia vernicata, Senecio sp., Charieis sp., Anthospermum æthiopicum, Euphorbia sp., Zygophyllum cordifolium, Justicia capensis?* at the rocks of Nus, in August—*Leucas sp., Asparagus sp., Oxalis sp.;* Kowsie river—*Monsonia pilosa, Lessertia sp., Charieis sp., Pyrethrum sp., Tanacetum sp., Mesembryanthemum sp., Hebenstretia sp., Heliophila sp., Osteospermum sp.?* bulbous plants.

MINERALOGICAL SPECIMENS.

I stated before, that the geology of the countries traversed on the expedition showed a primitive formation throughout. Of the copper ore submitted to Sir John Herschel at the Cape, he said, " that the specimen which he examined, in which was none of the madrix, appeared te be the submuriate, or rather, that particular chloride, which, on a calculation of its presumed atomic constitution, gave a result of 64 or 65 per cent. of copper."

The specimen assayed in London by Mr. Johnson, of Hatton Garden, gave 28, with 4 of sulphur; and perhaps the average of the Orange river copper is about a third, which is quite sufficient to induce a company (now forming) to work it.

Sir John Herschel read before the Literary and Scientific Institution of South Africa, of which he was President, the following notice of a chemical examination of a specimen of native iron, from the east bank of the Great Fish River, and which I had presented to the Institution.

" The specimen in question weighed originally 21·79 grains; 3·12 of which were separated, and submitted to a hasty preliminary examination for the detection of nickel, if any, but the quantity proving too small, the whole of the remainder was operated on in a subsequent trial.

" The iron was highly malleable and tenacious, and apparently of excellent quality, with a somewhat whiter and more silvery lustre than belongs to the metal in its ordinary state, and apparently little liable to oxidation, qualities which are observed in iron, of what is usually considered undoubted meteoric origin.

" I should not think it necessary to detail the steps of the analysis, by which the presence of nickel in the proportion of 4·61 per cent. was demonstrated, but for a peculiarity in one part of the process by which an inconvenience of frequent occurrence in chemical operations, and of a very embarrassing nature, was obviated, and which may prove useful as a hint to young analysts in other cases.

"18·67 grains of the iron in one piece, were digested in dilute nitric acid, which dissolved the whole with the exception of a trifling quantity of black scaly matter, apparently amounting to about a quarter of a grain. Towards the end of the solution the iron more than once brightened on the surface, and assumed that peculiar and singular state of resistence to the action of the acid which I have described in the Annalis de Chimie for September, 1833, and which has since been the subject of so much interesting discussion by Professor Schoenhein, Sir M. Farraday, and others. In consequence, it was necessary to apply and maintain heat to complete the solution.

" The nitric solution was evaporated to dryness, water added and evaporated a second and a third time. By this the whole of the iron was peroxidized, and nearly the whole separated. It was then diffused and boiled in water, to which a few drops of nitric acid were added, to take up any oxid of nickel which might have been deprived of its acid by over-heating, and set aside for subsidence, filtration being out of the question.

" After standing a week, however, it was still perfectly opaque, and loaded with suspended peroxide of iron, and to get rid of this was the next object.

" Lead being a metal easily eliminated, and incapable of interfering in any of the subsequent processes, its introduction seemed not likely to prove any source of further embarrassment; a few drops of dilute nitrate of lead were therefore added, and being well mixed, as much sulphuric acid as would saturate the lead, and a little more was added, and the whole boiled. The precipitation was complete, the lead carrying down with it all the suspended ferruginous matter, and leaving a clear liquid of a greenish hue, in which the presence of lead could not be detected.

" The remaining iron held in solution, was removed by heating it with excess of carbonate of lime, in the manner pointed out by me in the Phil. Trans. for 1821, when, after filtration, a liquid remained of that peculiar tint of pale green which characterizes the solutions of nickel, and of considerable indensity.

" The presence of this metal was ascertained on concentrating the solution by the usual tests, and its quantity concluded, viz.—0·86 grains, or 4·61 per cent. on the specimen analysed.

" Thus it appears that the specimen brought home by Capt. Alexander, has equal claim to a meteoric origin with any of those masses of native nickeliferous iron which have been found in different localities, and to which that origin has, without other evidence, been attributed.

" All those specimens, however, have, so far as I know, been insulated single masses. But what constitutes the peculiar and important feature of this discovery of Capt. Alexander, is the fact stated by him of the occurrence of masses of this native iron in abundance, scattered over the surface of a considerable tract of country. If a meteoric origin be attributed to all these, a shower of iron must have fallen, and as we can imagine no cause for the explosion of a mass of iron, and can hardly conceive a force capable of rending into fragments, a cold block of this very tenacious material, we must of necessity conclude it to have arrived in a state of fusion, and been scattered around by the assistance of the air or otherwise, in a melted, or at least, softened state."

Such, then, is a notice of some of the most interesting specimens of my collection of Natural History, which (besides several karosses, arms, and implements of the natives) was as large as I could make it, under the peculiar circumstances, explained in the preface, and elsewhere in the narrative, which served to confine it within portable bounds. Ere long, I expect that my follower, Taylor, who returned to Namaqua Land to collect, will enable me to lay before the public some more new and rare specimens.

APPENDIX No. II.

ITINERARY OF THE AFRICAN EXPEDITION OF DISCOVERY OF 1837.

CAPE TOWN TO WALVISCH BAY—NIAIS—CAPE TOWN.

CAPE TOWN to	Miles	LILY FOUNTAIN to	Miles
Laubscher's Farm	17	Keerom	40
Proctor's Farm	17	Komakas	20
Malmsbury	16	Bont Koe	33
M. Smit's Farm	20	'Kama	35
Berg River	17	Jackal-Holes	35
Hanekom's Farm	22	'Aris	15
Kruis	21	ORANGE RIVER MOUTH	22
Company's Drift	15	Lily Fountain	200
Lang Vley	15	F. Coetzers	12
CLANWILLIAM	24	Silver Fountain	27
Uitkomst	24	Copper Berg	21
Vley	15	Koekus	15
Outspan	20	Bezondermeid	24
Ebenezer	24	Eerebies Fountain	18
Olifant's R. Mouth	24	Henkrees	18
De Toit's Farm	20	Koran Ford	24
Kalk gat	18	Sand Fountain	24
Paddegat	30	Ahuries	12
Zwartdoorn River	27	NESBITT'S BATH	24
Groen River	18	Dry Plain	18
Hoog Fountain	21	Kururu	26
Horn gat	27	Outspan	21
LILY FOUNTAIN	4	Naros	18

Naros to	Miles	Bethany to	Miles
Outspan	24	Koanquip River	25
Africaner's Kraal	12	Tamuhap River	6
Dry River	20	'Nanees	9
Salt River	20	Dry Outspan	18
Dry Outspan	20	'Tuais	7
Nesbitt's Bath	44	Koanquip River	21
Dubbee 'Knabies	18	Dry Outspan	15
'Knabies	18	Dry River	19
Kurehas	6	Ansabib	16
Old Kraal	12	Great Fountain	22
'Kanus	10	Huntop River	22
Aribanees	16	Arigha 'oup River	21
Keikap	22	Arigha 'oup River	7
Kamopkams	12	'Oosep River	7
'Kanus	50	Narop River	10
Plain	9	Narop River	15
Height	10	Nameless River	16
Chubeechees	25	Bull's Mouth Pass	9
Plain	16	Chuntop River	7
Broken Liver Fountain	10	Nameless River	2
Dry River	22	Chuntop River	19
Nanebis	11	Ababies	12
'Kamop Ford	7	Dry Valley	30
Dry River	10	Desert Outspan	13
Great Fish River	14	Kuisip River	13
'Neims	10	Off-pack (on River)	8
Dry Outspan	9	Ditto	8
Habunap	13	Ditto	20
Dry Outspan	25	Ditto	12
Kusis River	10	Ditto	25
Bethany	22	Ditto	15

Kuisip River to	Miles
Off-pack (on River) .	16
Ditto . .	16
Red Bank . .	8
Huts . . .	13
Walvisch Bay .	13
Sand Down .	13
Red Bank . .	15
'Gnu'hooas . .	24
Gnei'tueip . .	8
'Hou'tous . .	20
Dry Off-pack .	35
'Humaris River .	28
Kuisip . .	17
Kereekama .	12
'Onakusis . .	20
'Abashouap .	13
'Unus . .	9
Chu'ntop . .	18
Taop River .	21
Chunchuap .	28
'Naraes . .	26
'Ni'ais . .	12
Keikurup River .	12
Glenelg's Bath .	26
Gnu'tuais . .	10
Tuap River .	18
Kukama . .	21
Chounp River .	30
Hoakosams .	20
Hatep . .	19
'Kuis . .	8

'Kuis to	Miles
Kaikum River . .	12
Kaikum River . .	11
'Nubapis . . .	21
Chup River .	28
Kuhap . .	12
Huntop River .	16
Huntop River .	7
Keisu River .	17
Dry Off-pack .	21
Kutip River .	30
Ku''ums River .	11
Koan'quip Rivers .	27
Lion Fountain .	20
Dry River . .	20
Henrick Buy's Place	13
Bethany . .	15
'Quahanap River .	7
Dry Outspan .	21
Hudap . .	9
'Ukanip River .	22
Outspan . .	5
'Hoons River .	27
'Hoons Fountain .	9
Kubieb River .	18
Heineip River .	24
Outspan . .	13
Heris . .	10
Valley of Scorpions .	24
Orange River .	24
Outspan . .	10
'Ubip . . .	16

'Ubip to	Miles	Meer Kastel to	Miles
Aneip	9	Droog Kraal	22
Rocks of Nus	15	Riem Hoogte	20
Kuras	30	OLIPHANT'S RIVER	5
'Kama	18	Puts Vley	20
Ukribip	18	Heerlodgement	10
Haruhap	21	Uitkomst	23
Kowsie River	12	Lang Vley	18
KOMAKAS	24	Stink Fountain	18
Plat Klip	18	Boomzager's Farm	9
Kok Fountain	14	Field-cornet Hanekom's	14
The Poort	15	Field-cornet Lambreght's	24
Klip Vley	18	Cobus Smit's (24 Rivers)	30
Skuin's Kraal	10	Vogel Vley	12
Moordenaar's Kraal	25	Eikeboom	15
Briuntjes Hoogte	6	Paarl Village	20
Kogel Fountain	13	Stellenbosch Village	16
Meer Kastel	14	CAPE TOWN	36

.*** The translations of most of the names in this Itineray are inserted in the Narrative of the Expedition.

APPENDIX No. III.

NOTES ON THE COLONY OF THE CAPE OF GOOD
HOPE.

1. The importance of the colony of the Cape of Good
Hope, in a military and a commercial point of view, to
Great Britain, is too well known to those acquainted with
our colonial empire, to need any elucidation here: that it
is an outwork to our valuable eastern possessions, and that
it has been to a certain extent, and that it might be to a
very great extent, an outlet for our manufactures and for
our redundant population, and that it might become an
emporium for a great South American trade, are sufficient
to induce attention to it, and a desire, in those who have
the power, to promote its prosperity.

2. A stranger who had never seen Cape Town before,
and who might have arrived there in the end of last year,
would immediately have concluded, after a hasty outside
view, that the aspect of affairs was very prosperous,
for he would have seen no ruins, but many new houses
rising in various directions. However, these appearances
of prosperity were deceitful—the influx of compensation
money for slaves had given a temporary show of wealth;
and as, unfortunately, there are hardly any other means
of investing money except in houses, there is, at present,
quite a rage for building.

3. But it is painful to record the fact, that at the beginning of this year a cloud rested on the Cape of Good Hope,

and that the colony was in a languishing condition. We made diligent inquiries into the causes of this, and we found, from the conversation of intelligent individuals, that the unpromising state of affairs there was thus accounted for: first, from the evils attending party spirit; and, secondly, from the extensive emigration of the Boors from the colony.

4. In this, as in other colonies and communities, there is a good deal of party spirit, and a want of combination for the general good. At the Cape, a set of men, disappointed in their ambitious views, had been constantly at work to thwart and oppose the colonial government: hence originated a spirit of party, which broke up the community into rival and conflicting interests, producing weakness and vacillation.

5. Thus, also, the Home Government, between conflicting statements, have not been able to ascertain what may be really the best plans for benefitting the Cape, which has consequently been left in a manner to itself. Soon, however, it is to be hoped, the aspect of affairs will improve there; and it is earnestly recommended, that the merchants and others interested in the Cape, will bestir themselves, and by the formation of companies for useful objects, such as for the purpose of introducing steam vessels on the southern coasts of Africa, to clear the mouths of rivers, to improve the communication between one part of the colony and another, &c., that the tide will turn in favour of this old and improveable colony.

6. The value of a colony to the mother country, as a means of assisting the revenue at home, and which can also pay its own expenses, is always an object of primary importance. Now, it is believed that a much greater revenue might

be derived from the Cape than there is at present, and without distressing the colonists, thus: the present duties are three per cent. on British imports, and ten on foreign goods, while by doing away with certain objectionable taxes, six and twelve per cent. might be charged on imports. There are no duties on Cape wine or tobacco, whilst at Sydney the duties on liquors are very properly high, and produce a great revenue. The merchants in England would, doubtless, object to the dues on their goods being raised at the Cape, but we must study the general good, not private interests.

7. St. Helena was formerly much resorted to for supplies, and is so still; but provisions there, from unavoidable circumstances (such as the old residents leaving the island when pensioned off by the East India Company on the transfer of the colony to Her Majesty's Government) are now dearer and scarcer than ever. Last year, seven hundred vessels touched there, most of which could not get supplies at all. Two years ago, property which was valued at 4000*l.*, sold, in October last, for 400*l.*! It should therefore stimulate the inhabitants of Cape Town to entice more ships to refresh there, seeing that they will only meet with disappointment if they put into St. Helena. We trust that, with an influx of emigrants, St. Helena will revive.

8. But now, also, provisions are so high at the Cape, that potatoes even, can be used by few respectable families; meal, which used to be twelve dollars (at 1*s.* 6*d.*) the muid, of two hundred pounds, is now thirty-five dollars; meat is 9*d.* the pound, butter 4*s.* 6*d.* ! The cause of the high price and scarceness of supplies, is in consequence of the emigration of the border farmers, of which hereafter; and as the ships

consumed 100,000*l.* worth of provisions, &c. annually, and now threaten not to touch at the Cape again, owing to high prices, the colony must therefore sustain a grievous loss.

9. A great inconvenience experienced at the Cape is in the landing of cargoes from ships: there is at present nothing but a miserable wooden jetty, quite inadequate, as to size, for the goods with which it is constantly crowded. The greater part of the cargoes cannot be landed on this jetty at all; the lighters approach the beach in its neighbourhood, and Mozambique porters then carry the goods through the water on their brawny shoulders. In summer, these poor men are roasted above and chilled below; and when the thermometer is under 60° they will not work at all. Strong though these men naturally are, they seldom live beyond thirty years with this sort of labour. It is gratifying, however, to know, that this wretched way of proceeding is soon to end, the local government having approved of a plan of a dwarf jetty submitted by the Surveyor-General, the commencement of which was ordered in April last.

10. About the year 1832 a substantial stone pier was commenced by Major Michell, near the Amsterdam battery, by order of the Home Government, the works of which were suspended in the following year (also by orders from home), although already advanced two hundred feet in length, and considerable masses of most valuable material collected near the site, for its continuation; thus rendering useless the expence incurred, and subjecting what was already achieved, to demolition by the elements. This was a work of great importance to the shipping, as affording means of rendering assistance to vessels in distress, and facilitating the landing of boats, whether the

wind blew from the N.W. or S.E. The site was selected, with great judgment, by the port captain (Captain Bance, R. N.) and highly approved of by Admiral Warren, after a personal inspection. It promised, indeed, every advantage short of *sheltering* vessels at anchor; the expediency of a work to effect which is questioned by engineers, as local peculiarities might endanger the shallowing-up of the bay. That the completion of the pier should have been arrested, is therefore a thing to be regretted.

11. Lighthouses are much required at Cape Receif (Algoa Bay), at Cape Lagullas, and at Cape Point; these, and the Cape and Algoa Bay piers, ought to be paid for by the general shipping interest, and the funds of the colony ought not to be employed in these works for the general commercial advantage.

12. The Surveyor-General recently found, at Cape Lagullas, a site formed by nature for a lighthouse, viz., a hill of tuff-like limestone, four hundred and fifty feet high, and seven hundred yards from the beach, and where, also, was abundance of stone, lime, sand, and water, on the spot. The cost of a lighthouse would therefore be trifling, with materials so conveniently situated for constructing it; and it would assuredly save hundreds of lives, and thousands of pounds.

13. In the course of last year (1837) accident afforded the inhabitants of Port Elizabeth a hope of obtaining what the low state of the public finances had until then refused them, a jetty of some sort whereon to land passengers and goods. The Feejee merchant vessel foundered at a distance of about two hundred yards from the shore, and firmly imbedded itself in the sand. An intelligent and enterprising young merchant (Mr. John Thornhill) im-

mediately conceived the project of turning the misfortune of this vessel to good account for the port, by using the wreck as a platform for driving the first system of piles for the construction of a wooden jetty, and uniting with other individuals, he purchased the hull of the Feejee, and immediately drove the first eight piles. A plan of a joint stock company was then prepared, and the consent of Government was obtained; a favourable report as to the probable success of the undertaking being made by the Surveyor-General, the shares were soon filled up, and the work is progressing towards its completion, thus verifying the homely saying, that "it is a bad wind that blows no one good."

14. Servants at the Cape, owing to the want of a stream of emigration from England, are generally very dear and very bad; thirty rix-dollars a month, is a common price for an ordinary black cook, and twenty-five for a groom; and to keep a servant six months is considered wonderful. A Member of Council said to me, one morning, "To-day I am without a servant in the house: though we gave our people good wages and victuals, there was no contenting them; they have all gone off through mere whim and caprice, and I dont know how I am to get my dinner dressed to-day. Some time ago I tried to give a little dinner to two or three friends from England, and when my wife retired after dinner, hearing a noise in the kitchen, she found that the cook had got drunk, and had gone off to Cape Town, leaving the remains of the dinner to be finished by the dogs!"

15. Another friend said, " I have got a drunken fellow of a cook, and often get my dinner at eight, which was ordered for six o'clock; and if I speak to the cook, he says

he'll go away; and its of no use turning him off, for I can get no better. Servants are so bad and insolent here, that housekeeping is quite intolerable."

16. The grand panacea for the Cape is *labour*. The climate is the finest in the world, I think, without any exception. The capabilities of the soil are great; and there is a constant demand for workmen of all kinds. The Cape wines and brandies are so attractive to the generality of mechanics, that not one in twenty can resist their seductive influence. Clever artisans exist a few years in a state of constant and wearying excitement from liquor, and are then transferred from the hospital to the churchyard; the hands are therefore few, and high priced.

17. The Malays are tolerable carpenters for rough work, but cannot execute cabinet work like Englishmen. Ship loads of excellent farm servants and mechanics are constantly passing the Cape for New Holland and Van Dieman's Land, but none stop here with their valuable living cargoes. All connected with the Cape in England must therefore exert themselves to introduce emigrants into the colony. Since 1820, when 5000 British settlers were sent to the Eastern province, no emigrants have been introduced into the colony. Before the Caffer war of 1835 the British settlers in Albany were in a very prosperous condition. Trade and agriculture flourished, the exports and imports at Algoa Bay were rapidly on the increase, when suddenly the prospects of the settlers were blighted by the destructive invasion of the Caffers. It is anticipated, however, that ere long matters will be so arranged in the Eastern province, that there will be a certain prospect of peace and tranquillity for years to come, when it will be advisable for settlers to proceed to Albany, and in the

mean time, at and near Cape Town, there is room enough and safety for them.

18. A depot should be established at Cape Town, where emigrants arriving from England could be kept till it was advisable for them to leave for the situations which offered themselves for them. They would thus be kept out of harm's way on first landing, and have a chance of being eventually better provided for.

19. Some time ago a society was formed in England for the purpose of conducting the emigration of orphan and other poor, *but not depraved*, children, from Great Britain to the colonies; four hundred and nineteen male-juvenile-emigrants, and ninety-five girls, have already been taken from the streets of London, where they were reared in the midst of vice and misery, and have been sent to the Cape of Good Hope.

20. The members of the society, who incur the great responsibility of sending these children from their native land, have been at great pains to ascertain the reception the children are likely to meet with in the colonies to which they may be sent, and accordingly required interrogatories to be satisfactorily answered, by those who could give the best information, regarding the demand for servants, and the prospect of their meeting with immediate employment from persons of good character, before sending abroad, with the sanction of Government, the children under their care, so as not to render emigration a mere change of place, without benefitting the condition of the emigrant youth.

21. The average rate of wages at the Cape for a boy of fourteen years of age, engaged in agricultural pursuits, has been about 3*l.* 10*s.* per annum, on a seven years' ap-

prenticeship, with good food and clothing. Artisans give about the same; but the demand for boys by artisans is less than by farmers. Indoor servants receive about 4*l.* per annum.

22. Boys are usually bound by indenture, and generally for seven years, or from five to seven years—also a contract legally entered into in Great Britain, for services to be performed at the Cape, will be enforced in the colony.

23. There is a constant demand in Cape Town for the labour of boys as house servants, grooms, or gardeners. But the supply of this labour from home should of course be regulated by constant advice from the colony, and which is now given by a Cape society (for the management of juvenile emigrants) which communicates with the London society.

24. By the Cape society it is recommended that boys from eleven to fourteen years of age should be sent out; that with regard to girls, as indoor servants, at first they are worth little more than their food and clothing, but after their seventeenth year their wages vary from 8*l.* to 30*l.* per annum, according to quality.

25. No girls, it is understood, have been employed in trades at the Cape; but it is supposed that if they were thus employed, their wages would amount to from 2*l.* 10*s.* to 3*l.* 5*s.* per annum. There is a great demand for girls as indoor servants; and the proper age for their being sent to the Cape, appears to be from ten to twelve.

26. The boys who were first sent out had their passage paid by the masters to whom they were indentured, at the Cape, and they received the first year, their food, clothing, and washing, besides sixpence or ninepence per week as

pocket money, two-thirds of which was invested in the Savings Bank for their benefit.

27. That a number of the juvenile emigrants will turn out well, there is little doubt, particularly those who have not been exposed to much contamination at home, and who fall into good hands at the Cape—others, again, appear quite incorrigible, the vices of lying and stealing seem to be so implanted in them—whilst a third class fall into the hands of such bad masters, where they are allowed to herd with Hottentot servants, whose general habits are very demoralized, that it is a question, whether they had not better have remained at home, even with the chance of transportation to Australia, than to have been sent to the Cape, to degrade the character of white men, in the capacity of *white slaves*, as the Dutch call these apprentices.

28. A little work, by the philanthropic Captain Brenton, R. N. (an active member of the London committee of juvenile emigrants) points out clearly and distinctly the advantages of sending juvenile emigrants to the colonies.

29. In the matter of Cape improvements we hear a good deal, ever and anon, about removing the bars of rivers—about canals to connect one river with another—and of a magnificent project to unite the waters of Table and False Bay, to avoid the doubling of the Cape itself. These schemes may some day be attempted; but before they are, I think that the first object of Government, ought to be to improve the present landing places at Table and Algoa Bay, by building piers, and laying down moorings—the purposed intercourse by steam between the two Bays cannot succeed, unless passengers and goods can be landed at all times, and with safety, on good piers.

30. Some maintain that the English and Dutch will

never amalgamate at the Cape, and will never "pull together" for the general good; but this I hope is a mistake. Certainly, a better feeling would have existed between the English and Dutch than what exists at present,
if the measure of the English philanthropists of sudden
emancipation of slaves had not taken place. The Dutch
were unprepared for this act of justice to their dark
dependants; it took them by surprise; and it is not to be
wondered at if they considered themselves hardly dealt
with, though it is to be hoped that time will heal the
wounds which they consider have been inflicted on them.
They maintain that they have reason for complaint,
since a great many of them only got about one third of
the amount they paid for their slaves, and thus, as was
before noticed, a farmer who gave, a few years ago, 800*l.*
for slaves to cultivate his farm, receives now only 300*l.*
compensation for the loss of his people; but it was difficult to prevent cases of hardship, like this occurring in carrying through the great and most humane measure of slave
emancipation. How the orphans of slaves, the sick and
the aged are to be cared for, and how the farms are to be
cultivated in 1839, after the expiration of the apprentice-
ships in 1838, and when the late slaves will flock to the
towns, it is difficult at present to say; but it is believed
that Government has duly cared for all this.

31. The Dutch, of late years, have adopted many of the
customs of the English. Thus, the late bare and var-
nished floor is covered with a carpet; and the Dutch
houses generally are better furnished than they used to be.
No farmer comes to market, without being respectably
dressed in English broadcloth, with a good hat on his
head, and polished boots or shoes on his feet. Skin

trowsers and untanned *field shoes* are seldom now seen at the Cape.

32. But the irrigation of land has been almost entirely overlooked and neglected by the Dutch farmers. Contaminated by slavery, they will seldom take off their coats to work, and neither dig wells, nor construct dams. I do not remember to have seen a single well beyond the precincts of Cape Town and Graham's Town, and hardly anywhere a dam, even for watering a garden. The attention of emigrants ought to be particularly directed to assist nature by collecting the water in the rainy season.

33. At present there is in agitation the formation of a good road between Stellenbosch and Cape Town, thirty-six miles, across the sandy flats which so materially impede the transport of produce to market. This road would be a most important improvement, and ought not to be lost sight of. Generally throughout the colony, the want of navigable rivers ought to be supplied with good roads and bridges. The first and best means to improve any country are, to afford facilities for communicating between one district and another, and to enable produce to be easily transported to a market. The formation of the excellent roads over Sir Lowry Cole's Pass, and Houw Hoek Pass, by the present Surveyor-General, have been very advantageous for the colony, and of great benefit to the farmers using them; but these great improvements ought to be followed up by a hard road over the sands to Stellenbosch, for the old *track* is in many places axle deep, and the sides of it are strewn with the bones of dead horses and bullocks, which have here miserably perished. The earnest cravings of the Cape people for the blessing of a good road to Stellenbosch, induced them, last year, to go

so far as to subscribe (in the form of shares) to the amount of seventeen or eighteen thousand pounds, towards the expenses of the work, stipulating for the turnpike dues for reimbursement; but the Surveyor-General's estimate amounted to 28,000*l.*, a sum not to be compassed in the above manner, and so the hope of having a hard road over the " Flats and Downs," has for the present, at least, disappeared. The bold improvements required in the way of roads can scarcely be expected to succeed under any private body; but in time Government will doubtless forward such undertakings.

34. Opening up a country by roads ought to be followed by opening the minds of the people by education. In the country districts of the Cape colony there is a great want of good teachers. A normal school ought to be instituted without delay at Cape Town, to prepare masters for country charges. The rising generation of Dutch would then see the great advantages of an improved manufacture of wine; of attending to the improvement of the fleeces of their flocks; of irrigating the land; of taking every advantage of their situation, and of developing the capabilities of this valuable, but generally unimproved, colony.

35. The difficulty of inducing colonists to invest their idle money in useful and productive undertakings, such as turnpike roads, bridges, banks, fisheries, wharfs, railway and insurance companies, or other beneficial stocks, arises, in a great measure, from their being unacquainted with the generally improving productiveness of capital, employed in all such necessary and beneficial works. Persons who do not see the good security of capital laid out in banks, turnpike roads, wharfs, and other public and useful

working companies, are generally fearful of investing money in such undertakings; but these, in most countries, and particularly in new countries, are generally the most secure investments, and produce greater returns of interest, when prudently and properly managed, than trade or commerce.

36. It is hoped, however, that the few public establishments which have been lately formed in the colony will give the colonists some experience in these matters, and draw the idle and unproductive capital from its slumbers in their *kists*, into active use in public improvements. We have heard that the South African Insurance Company's shares of 10*l.* paid up, are now worth 50*l.*; a gain, in six years, of about 40*l.* on each 100*l.* share; that the shares of the Cape of Good Hope Insurance Company, established only about two years ago, have gained about 50 per cent. That the new joint stock bank just established, called the Cape of Good Hope Bank, is selling its shares at an advance of 5*l.* on a share of 50*l.*, and that the splendid road over the Hottentot's Holland Mountains, made in the time of His Excellency Sir Lowry Cole, a late governor, and worthily named after His Excellency, who zealously promoted it, is now giving a return of 15 per cent. per annum to the Colonial Government for the capital expended in making it. This is a fine example to the colonists of what capital will produce in roads, and what solid security there is for capital invested in such undertakings.

37. It is lamentable to see the state of the fisheries at the Cape. Here, along the shores, and in the bays particularly, both Table and Simon's Bay, at the proper season, are numerous whales, ready at hand for the harpoon; but, from the inexperience of those engaged in the trade, and the want of some further and more precise regulations for

the fisheries, not one half of the valuable fish which appear in these bays are taken. The Colonial Government might, perhaps, make a considerable addition to the present revenue out of these fisheries, if some alterations could be made in the British fishery laws in favour of this colony; for the strictness and peculiarity of the British fishery laws render whale fishing at the Cape a very precarious business. The strict duty which the British fishing code compels the officers of Her Majesty's Customs to carry into effect at the Cape of Good Hope, particularly in the proofs to be given of the oil being *British*, acts as a great discouragement to the colonists entering into the whale fishing trade. If a single foreigner happens to be in a boat's crew which takes a whale, the oil is not British; if a boat is not British built, or a whale line, or whale spear, or indeed, if there is any foreign article used in the killing of a whale, the oil is not only classed as foreign, but we have heard that it is liable to confiscation. Nowonder our British fisheries in this part of the world do not flourish or succeed; no wonder we see foreigners of several nations, and particularly the Americans, resort to our coasts, and take out of the bays, and from the islands of South Africa, thousands of tons of oil, and thousands of seal skins annually.

38. The British government should look to this; for the Crown loses a large revenue, and the country a profitable and extensive trade for the lucrative employment of its ships and seamen. We fear that we shall never have experienced seamen for the southern fisheries while our laws are so very particular, and this valuable trade is so neglected by the British nation. Why should the Greenland fisheries be so protected and cherished, and those on the African coasts and bays be so neglected?

39. American and other foreign whaling ships visit Table and Simon's Bays for refreshments to the extent of many tons annually, and then show the extent of the trade, and the great value of their cargoes of sperm and black oil and bone.

40. The new bank at the Cape is succeeding beyond calculation, and has the full confidence of the public; and there is little doubt, but that, in the course of next year, the shares (paid) 33*l.* 6*s.* 8*d.*, will be worth 50*l.* It is, as has been already mentioned, a Joint Stock Bank, begun on the 1st of August on a trust deed. The Home Government refused at first to sanction the ordinance passed by the Colonial Legislative Council for its formation, under the plea that it would reduce the colonial revenue, by taking away part of the revenue of the Government Bank, the profits of which go into the Colonial Treasury. But now Her Majesty's Government having been induced wisely to sanction the establishment of the new bank, it has already, by competition with the Government bank, done service to the colonists by causing this last to abandon its system of imposing stamps on the bills that it discounted, and causing it to lower its rates of discount from 6 to 5 per cent., and this in fair competition; and as this Joint Stock Bank is upheld by colonists, who belong to the principal mercantile houses in the colony, and by their partners in England, there is no doubt that in the course of a short time it will take away the principal revenue of discounts which the Government Discount Bank had in the colony, and will also prevent any tampering with and deranging the colonial currency, which had been so long and so often agitated.

41. For many years there has been only one bank in the whole of South Africa, and that was the Government

bank, which issued a local paper currency, varying so much in exchange with British sterling money, as well as in its exchange with the currencies of other countries, that the original rix-dollar, which was issued at 4*s.* sterling, in some measure nominal, fell in the progress of time to the sterling value of 1*s.* 4*d.*, and is now fixed by the British Government at 1*s.*6*d.* The Commissariat receiving the rix-dollars now at that rate for bills on the British Treasury, London, less 1½ per cent. premium.

42. At present there are no other banks in South Africa except those at Cape Town, but the New Joint Stock Bank intends to form branches in the principal towns of the colony, by which trade and commerce, and particularly agriculture, will be much convenienced.

43. The capital now flowing into the colony, will, we think, produce some effect on the wine, and render it more acceptable to the English taste. Were it not for the colonies of Australia, it would be impossible for the Cape merchants to dispose of the inferior wines, and they would be under the necessity of being more particular, in order to compete with the Portugal and Mediterranean wines, were it not for the outlet to New South Wales. Unless some great improvement speedily takes place in the quality of Cap wines, and which is quite possible with care, the Cape must still look to Australia as the place for the chief reception of its vinous produce.

44. Wool is succeeding as well as can be expected with a thin population, and it is not thought that New South Wales has done better than the Cape, when we take into consideration the state the Boors were in up to the time the British settlers arrived, cut off from all enterprise in such a manner as to prevent useful combination and energy.

45. A Mr. Reitz, of the firm of Reitz, Breda, and Jou-

bert, who superintends the largest flock in the western province, has, we believe, fifteen thousand pure Merinos in the district of Swellendam. Mr. Reitz is an intelligent and enterprising young man.

46. The charge of indifference to the improvement of wool, against all the farmers of the colony, I find is not so well founded as would at first sight appear, and a comparison with New South Wales is unfair. The Cape agriculturists had already invested the greatest portion of their capital in vineyards, when the Merino sheep came into vogue; the New South Wales colonists had to begin *something*, and fortunately hit upon the right article, or it might rather be said were forced to it, as neither wine nor grain would answer in Australia at all.

47. The most of the Cape farmers were poor, and therefore prudent in not risking a new experiment. The New South Wales people had made a little money at first by selling, it is said, rum at one pound per bottle. The English settlers of Albany, enterprising and intelligent as many of them were, did not commence sheep farming immediately after their arrival in the Cape colony; even the great wool grower, Lieut. Daniell, R.N., had so far involved himself with agricultural speculations, that he was at one time on the eve of embarking for England. At first it was not at all certain that woolled sheep would answer in the Cape colony. The Messrs. Cloete, Vander Byl, Myburgh, &c. who have now large flocks of woolled sheep, derived their money in the first instance from vineyards, or they might have been much further advanced than they are at present, since they had to contend with the disadvantage of slave labour, always indifferent and expensive. Other old farmers, as the Van Reenens, the Bredas, Eksteens, &c., are not careless about improvements in sheep

or any thing else.   In the year ending 5th January, 1837, there was exported from the Cape colony 116,574 pounds of wool, the value of which was 7353*l.*

48. Having had occasion to mention the name of Lieutenant Daniell, we feel bound to record, to his great credit, that he has lately appropriated a valuable section of his farm for the site of a church (in connexion with the Church of England) and parsonage ; towards defraying the building expenses of which, he and other members of his own family have subscribed a sum equal, at least, to half the entire cost.   The designs for the above were furnished by the Surveyor-General.

49. The abandonment of the Cape colony by a large body of Dutch colonists within the last two years, and their withdrawing themselves from the protection of the English Government, is a most remarkable fact in the history of colonies.   We shall now shortly state the reasons these colonists assigned for leaving their farms, where most of them had enjoyed all the comforts suited to their condition, and where their forefathers had dwelt before them, and at once plunged into a wilderness teeming with dangers, and where they must have known they should require to exercise a constant state of watchfulness against savage neighbours.   .

50. The emigrant Boors said, that the sudden emancipation of their slaves without adequate compensation for them, was one reason for their leaving the colony; and they carried a number of their slaves with them into the interior.   They were also vexed at the appointment of special magistrates to see justice done to master and servant ; for, they said, that the magistrates were too much inclined to listen to the frivolous complaints of the slave against his master ; and to fine a white man for

lifting his hand to a Hottentot, irritated the Boors beyond measure.

51. One discontented farmer said in my hearing, though, of course, I do not at all agree in the justice of his observations, " What can be worse than this? If I give a kloppé (klopje, or a little blow) to a slave, he immediately runs off to a magistrate and complains, and I am sent for from the middle of my harvest work, perhaps, and am obliged to ride twenty or thirty hours, to answer the complaint, and I come away, leaving five pounds as a fine. When I come home, I cannot help giving the Hottentot another kloppé, when I am fined ten pounds. And then, to make up for taking away our black slaves, the English have made slaves of their own children, and send them out here (the juvenile emigrants). What can be more disgraceful than this in Christian men! I shall sell my place, and be off to Natal."

52. Again the Boors complained that vagrancy was not suppressed, particularly in the border districts; that the country swarmed with coloured vagabonds of all sorts, who lived by plunder.

53. And, lastly, they said that many of them had been ruined by the Caffers, and they were afraid to remain in the vicinity of such dangerous neighbours, from whom they did not think they were sufficiently protected by a line of posts, principally occupied by Hottentots along the Fish River Bush.

54. These were the reasons the Boors assigned for leaving the colony; but there were others, also, which made old and young of both sexes *mad* to locate themselves beyond the borders. A few farmers had visited the country about Natal, where some Englishmen and others had esta-

blished themselves for the purpose of collecting ivory and buffalo hides, and who hoped to found a colony there, under the protection of the British Government. The Boors who visited this part of the coast were struck with its superiority over the Cape colony; for it abounded in water, grass, wood, and game, and the climate was also excellent. Now, though the climate of the Cape cannot be surpassed, yet there is sometimes a scarcity of rain, and consequently the pasture is not generally so abundant as it is at Natal, which is placed on the verge of the Indian monsoon.

55. The farmers on their return home inflamed the minds of their neighbours with a desire to visit the terrestrial paradise about Natal, and the adjoining territory of the Zoolahs; and, as they were afraid to go in a small body amongst some of the most warlike tribes of South Africa, they tried to get up an extensive emigration for mutual protection; accordingly, emissaries went through the Cape colony exciting the Boors against the English Government, and drawing a highly coloured picture of the region to the N.E. of the colony.

56. " There," they said, " was the garden of Eden, where Adam and Eve dwelt; every game animal is to be found there; and the trees are loaded with fruit of the largest and best description. Among other things of an extraordinary nature which are to be found there, the potatoes are so large that it requires a span of twelve oxen to drag off one of them !"

57. A movement took place not only along the eastern frontier, but generally throughout the colony; the mania for emigration rapidly spread, valuable farms of 6000 acres, with dwelling-houses, outbuildings, gardens, &c.,

were sold for the trifling sum of 200*l.* or 250*l.*, and were bought by merchants and shopkeepers in Graham's Town on speculation: one merchant, I know, bought three for the above sums. But some farms on the north-eastern frontier, I heard, were parted with for a new waggon, about 60*l.* in value; and, seemingly without regret, old and young, the aged and infirm, with their flocks and herds, moved off towards the wilderness.

58. The first party which emigrated in 1836, consisted of thirty families, and moved towards Natal, but, misled by the mountains, they passed it a considerable distance before they discovered their error; they then remained between Delagoa Bay and Natal. They were followed by other parties, some of which did not seem to have any intention of approaching the Indian Ocean at all, but keeping in the interior, they fell amongst the warlike Zoolahs of Moselekatsi, and as it was impossible to keep close together on account of the sheep and cattle, they were savagely butchered in detail by the Zoolahs.

59. But this was not to be wondered at, even though the Boors conducted themselves quietly. The Zoolahs are composed of two large tribes: one under Dingaan, whose usual residence is 170 miles N.W. of Natal; and another, under Moselekatsi, whose country lies between that of Dingaan and the town of Latakoo. The Zoolahs, like the Kamaka Damaras, use a single stabbing spear, rush to a hand to hand conflict, in the most determined manner, and are the terror of the neighbouring nations. They inhabit a fine country; still, as they are a great people, they find they have no more pasture or water than what they require for themselves, seeing, then, large parties of white men coming among them, and like a flight

of locusts devouring every green thing with their flocks and herds, and sitting down at the pools, the Zoolahs, to prevent a famine, destroyed the Boors when they had them at advantage.

60. A party of Boors, who had been worsted by the Zoolahs of Moselekatsi, sought refuge at Thabu Unchu, the station of a missionary, who had formerly laboured with energy in Namaqua land, the Rev. Mr. Archbell.

61. But the Boors did not long remain at rest, and being joined by others, they set out under a leader of the name of Maritz, to revenge themselves on the Zoolahs, and to regain what they had lost. Surprising their enemies at the town of Mosega, they slew three or four hundred of them, and recovered seven thousand head of cattle, and some waggons.

62. After the account of this victory had reached the colony, those who were before undecided to move, immediately packed up to follow their friends in the interior, and the tide of emigration set stronger than ever across the colonial border, though the government authorities exerted themselves to the utmost to stop it. Up to the 1st of April last, it is supposed that, including men, women, and children, not less than *twenty thousand white people* had gone across the borders, and had abandoned their native land for ever. This is a very serious loss to the population of a colony which only averaged one soul for every square mile, or *one hundred and fifty thousand* in all.

63. The last accounts from the farmers were still more disastrous than those which first reached the colony of their misfortunes in the interior, they had already suffered from the Zoolahs of Moselekatsi, and they were now fated to become the victims of the people of Dingaan also.

64. The cold-blooded massacre of a large party of the
Boors, by the orders of Dingaan, took place in February
last. The tyrant acted towards them with the blackest
perfidy: they incautiously reposed confidence in him,
trusted to the generosity of savages, and exposed themselves, unarmed and defenceless, amongst them, relying on
their professions of friendship.

65. It appears that another leader, Retief, a worthy man,
(and who had assisted the English settlers who came to
Albany in 1820, to the utmost of his power), in order to
ingratiate himself with Dingaan, had captured seven hundred head of cattle and some horses for him, from a chief
with whom the "Great Black One" was at war; and Retief
then went, with sixty mounted Boors, and forty Hottentots,
to Dingaan, on the 3rd of February, to deliver up what he
had secured for him, and to treat about land.

66. Apparently every thing was arranged in a satisfactory
manner with the tyrant, who granted the Boors whatever
they wanted; and they were overjoyed at their success
with him, when, on the 6th of February, as they were
preparing to saddle up and return to their families, encamped at some distance from Dingaan's residence, the
chief invited them to a dance and to drink milk, but he
asked them not to bring their arms within the royal kraal.
The Boors complied with the desire of the perfidious chief,
and whilst they were drinking, about two thousand of the
Zoolah warriors were ordered to dance before them; they
advanced and retired, and then forming themselves into a
half circle, at a signal from Dingaan they fell on the Boors
and their servants, half a dozen on each man, and dragging
them across a river to the "Hill of execution," they there
either twisted their necks or knocked out their brains with
knobbed sticks.

67. A missionary (the Rev. Mr. Owen) living at Dingaan's residence, was spared, and permitted to go to Natal, on giving up a waggon; and he was told that Dingaan had killed the Boors because he knew that they intended to kill him! Not content with what he had already done, the monster, immediately after the above tragedy, sent a strong force to surprise the Boor's camp, where about three hundred more fell, men, women, and children. A handful of men rallying after their first surprise, slew a number of the Zoolahs, who then retired.

68. As soon as the news of these disasters reached Natal, the white people there, English and Dutch, immediately assembled their people, and white and black went out, one thousand strong, against Dingaan, but he had fled with his hordes, and the Natal force returned without being able to effect what they intended.

69. Thus had the path of the deluded self-exiles been marked with blood, and it must so continue to be stained, either with their own blood or with that of the natives, whose domains they invaded, unless the tyrant Dingaan soon falls, or they retrace their steps to the colony.— Bitterly must the survivors of the unfortunate emigrants lament their precipitation in abandoning hastily a colony where, though many of them had suffered from the Caffers, they were at least more secure than among the savage Zoolahs, and where they might have expected that in time, the measures of Government would have placed the persons and property of those on the immediate frontier in security —whilst those at some distance from the Caffers had nothing to dread, and certainly had no excuse for taking the road to the Wilderness.

70. The character of the colonists of South Africa is not understood in England, as the whole of them are usually

accounted as men who are continually oppressing the natives, and are consequently retaliated upon; but to class
them all under one head, indiscriminately, is quite an erro-
neous way of viewing them—they ought to be divided into
three distinct classes, 1st. the Dutch cattle farmers, of the
north and north-eastern borders, and who resembled the
Back-Woods-Men of America, because they were brought
by their position in contact with the Boschmans, &c., and
between whom and the natives there was frequent warfare,
and, doubtless, many cruelties committed on both sides.
2d. The inhabitants of towns, and the Dutch farmers living
at a distance from the frontiers, who are peaceable and
well-disposed, and who never uplifted an arm against a
native, except to chastise a servant with a rod. And, 3d.
the English settlers in Albany, who were divided from
the Caffers by the military, and who never had it in their
power to plunder the natives beyond the colonial borders,
who outnumbered the settlers as ten to one. The settlers
were principally occupied with agriculture, and some of
them had trading stations among the Caffers. They (along
with many of the Dutch farmers) suffered severely by the
Caffer invasion of 1835; indeed they were the chief
sufferers.

Now, to conclude these rough notes, it is confidently
anticipated that the suppression of vagrancy, the promotion of emigration, &c., and an increased force of *English*
and native dragoons, or of soldiers who will act indifferently
on horseback or on foot (and a large proportion of whom
being our countrymen, could not be tampered with by the
natives beyond the borders), will secure the tranquillity,
and ensure the prosperity of the Cape colony generally.

" Viret in æternum."

J. E. A.

A SUMMARY VIEW of the TRADE and NAVIGATION of the Colony of the CAPE OF GOOD HOPE, from the 6th of January to the 5th of July, 1837.

	Vessels Inward.		Vessels Outward.		Customs Duties alone	Total Revenue.
	No.	Tons.	No.	Tons.		
					L. s. d.	*L. s. d.*
Cape Town -	189	69798	191	70202	16823 12 10	19052 14 0
Coastwise -	43	2985	43	3199	- - -	- - -
Port Elizabeth -	13	1996	11	1653	1794 3 10	1873 9 1
Coastwise -	19	2143	24	2576	- - -	- - -
Simon's Town -	22	6861	22	7250	139 10 4	226 0 3
Coastwise -	-	- - -	- -	- - -	- - -	- - -
Total, Colony -	224	78655	224	79105	18757 7 0	21152 3 4
Coastwise	62	5128	67	5695		
Grand Total -	286	83783	291	84900		

	Total Value of Imports.	Total Value of Exports.
	L. s. d.	*L. s. d.*
Cape Town - - -	508740 18 4	193263 5 0
Port Elizabeth - - -	53957 0 0	19180 0 0
Simon's Town - - -	1411 0 0	25 0 0
Total, Colony - -	564108 18 4	212468 5 0

THE END.

PRINTED BY WILLIAM WILCOCKSON, ROLLS BUILDINGS, FETTER LANE.

Lightning Source UK Ltd.
Milton Keynes UK
UKHW02f2102220418

321472UK00006B/74/P